# Tarantulas of Belize

**Steven B. Reichling**

**KRIEGER PUBLISHING COMPANY**
Malabar, Florida
2003

Original Edition 2003

Printed and Published by
KRIEGER PUBLISHING COMPANY
KRIEGER DRIVE
MALABAR, FLORIDA 32950

Copyright © 2003 by Krieger Publishing Company

All rights reserved. No part of this book may be reproduced in any form or by any means, electronic or mechanical, including information storage and retrieval systems without permission in writing from the publisher.
*No liability is assumed with respect to the use of the information contained herein.*
Printed in the United States of America.

---

**FROM A DECLARATION OF PRINCIPLES JOINTLY ADOPTED BY A COMMITTEE OF THE AMERICAN BAR ASSOCIATION AND A COMMITTEE OF PUBLISHERS:**
This publication is designed to provide accurate and authoritative information in regard to the subject matter covered. It is sold with the understanding that the publisher is not engaged in rendering legal, accounting, or other professional service. If legal advice or other expert assistance is required, the services of a competent professional person should be sought.

---

Library of Congress Cataloging-in-Publication Data

Reichling, Steven B., 1957–
    Tarantulas of Belize / Steven B. Reichling.— Original ed.
    p. cm.
    Includes bibliographical references (p. ).
    ISBN 1-57524-206-0 (alk. paper).— ISBN 1-57524-228-1 (pbk. : alk. paper)
    1. Tarantulas—Belize.  I. Title.

QL458.42.T5 R45 2003
545.4'4—dc 21     2002029847

10 9 8 7 6 5 4 3 2

Dedicated to
Ann

# Tarantulas of Belize

# Contents

| | |
|---|---|
| Preface | ix |
| Acknowledgments | xi |
| Chapter 1 A Tarantula Hunter's Diary | 1 |
| Chapter 2 Physiography and History | 11 |
| Chapter 3 The Tarantula in Context | 15 |
| Chapter 4 Why Care About Tarantulas? | 19 |
| Chapter 5 Why Worry About Tarantulas? | 23 |
| Chapter 6 Tarantula Habitats | 29 |
| Chapter 7 The Hidden Life of Belizean Tarantulas | 37 |
| Chapter 8 How to Find Tarantulas | 59 |
| Chapter 9 Collecting and the Law | 69 |
| Chapter 10 How to Identify Belizean Tarantulas | 73 |
| Key to the Tarantulas of Belize | 81 |
| Chapter 11 Species Accounts | 83 |
|     Pygmy Tarantulas | 85 |
|         Pygmy Tarantula, *Reichlingia annae* | 86 |
|         Highland Pygmy Tarantula, *Reichlingia* sp. | 88 |
|     Typical Tarantulas | 91 |
|         Mexican Redrump Tarantula, *Brachypelma vagans* | 91 |
|         Livingston Tarantula, *Citharacanthus livingstoni* | 94 |
|         Cayo Tarantula, *Citharacanthus meermani* | 96 |
|         Cinnamon Tarantula, *Crassicrus lamanai* | 98 |
|         Gutzke's Tarantula, *Metriopelma gutzkei* | 100 |
| Plates (11.1–11.14) follow page | 100 |

    Arboreal Tarantulas ................................................................. 103
        Maya Tarantula, *Psalmopoeus maya* .................................. 104
        Costa Rican Orangemouth Tarantula, *Psalmopoeus reduncus* ........... 106
Chapter 12 Extralimital Species ............................................................ 109
Chapter 13 Beyond Belize .................................................................... 113
Gazetteer ........................................................................................ 119
Bibliography .................................................................................... 121
Glossary ......................................................................................... 123
Index ............................................................................................. 125

# Preface

This book is intended to serve as a complete overview of the tarantula spiders of Belize and adjacent areas in the Yucatán Peninsula, and is written for anyone with an interest in tropical wildlife. Our understanding of these secretive creatures is far from complete, which will be obvious when the reader notes that seven of the nine species found in Belize were not described until 1996 or thereafter. For most species we know only the most basic aspects of their biology, and for some species we know virtually nothing about the way they live. However, Belize has now been thoroughly surveyed for tarantulas and I doubt that any undescribed species remain. Enough is now known about the tarantulas of this Central American country to draw an informative, if still unfinished, picture of their natural history and so I have written this book as a compilation of what is currently known.

As with many books written about some portion of the wildlife of the tropics, I hope that one of its principal accomplishments is to serve as a foundation for future biologists, both professional and amateur, upon which to add their observations and discoveries. When dealing with a completely unknown subject it's often difficult to know where to begin to unravel the mystery, and the task may seem too daunting to attempt. Perhaps this book will provide enough factual structure to encourage others to add their insights.

It is also my hope that this book helps to popularize a needlessly obscure group of animals. Only a few decades ago, books describing the amphibians, reptiles, and mammals of a region were odd concepts, and were intended to be read only by specialists in the field. Only birds, with their legions of dedicated amateur students, have been routinely served by regional field guides. This book marks the first attempt to present the tarantula fauna of a particular region in a way that is accessible and hopefully, interesting, to anyone fond of nature. This book was not written to keep the study of neotropical tarantulas an esoteric pursuit. This book was written to move arcane and dusty old facts out into the light and to make things which are currently known by only a handful of scientists now readily available to any naturalist.

In recent years there has been an explosion of information about tarantulas, appearing in books, magazines, and the internet. This has been almost entirely attributable to the growing number of people keeping these spiders as pets. For many of these individuals, their interest in tarantulas goes far beyond the limits of typical pet ownership, and many hobbyists are studying their animals in an effort to improve husbandry and increase breeding success. As a result, the source of most of this flourish of new information traces back to a captive spider. While this trend is welcome, it tends to neglect the natural biology and ecological role of the spider. Much has been written about tarantulas, but very little of it considers them from the standpoint of a wild animal. Yet there is an urgent need for such information as the tropical regions where most of these spiders live follow a rapid and seemingly irreversible trend toward destruction. Before we can decide how best to protect tarantulas in their native habitat, we need to understand the ecological role they play and how they live, what environmental factors impact them and which ones they are oblivious to. With this guide in hand, the reader can venture into the forests and savannas of Belize to find tarantulas and uncover the many secrets they still hold with a better prospect for success.

I have tried to minimize technical details that would only interest an arachnologist, and have written this book to be useful to anyone with a curiosity about neotropical tarantulas. Identifying tarantula species can be difficult for the nonspecialist because most of the descriptive literature is very old and obscure and usually so weighted down with technical terms and anatomical trivia that even the tarantula biologist finds reading it a tedious chore. With this in mind, I've followed the example of the reptile and amphibian field guides that I grew up reading, and that made herpetology accessible to so many amateur naturalists and encouraged the next generation of herpetologists. The identification key and accompanying text descriptions are written under the assumption that the reader will not have immediate access to dissecting scopes and digital calipers, and in most cases a confident species identification can be made with nothing more than a magnifying glass and this book. I've intended this book to serve as a field guide as well as a reference source.

A colleague once said that tarantulas are boring. I don't believe they have to be. They may *seem* boring because so much of their life takes place underground, in their burrows, where we cannot watch. If dug up and brought to the surface for study, a tarantula becomes no more accurate a representation of its life than a fish caught on a rod and reel and left flopping on the floorboards of a boat, and may seem as uninteresting. To be appreciated for what they are tarantulas have to be understood, and this takes patience and perseverance. I hope this book will encourage you to make the effort.

# Acknowledgments

My introduction to Belize began in 1994 with a serendipitous visit to Lamanai Outpost Lodge. That first visit introduced me to the remarkable Howells family: Colin, Ellen, Mark, and Monique. Without their generosity and kindness I would never have been able to pursue the work which resulted in this book. For their unfailing support and friendship I will always be grateful to my dearest friends, Mark and Monique.

Many other friends and acquaintances in Belize have helped me along the way. I am grateful to the Belize Forest Department, Conservation Division, Ministry of Natural Resources, particularly Emil Cano, Angel Chun, Earl Green, Natalie Rosado, and Raphael Manzanero for granting me the privilege of studying and collecting in Belize. Carol Farneti Foster shared many valuable collection localities and saved me much leg work by doing so. I thank the naturalists at Lamanai Outpost Lodge: Carlos Godoy, Benjamin Cruz, and Blanca Manzanilla for sharing their masterful knowledge of Belizean wildlife, and the rest of the Lamanai staff for their friendship and kindness. Bart and Suzi Mickler's hospitality while I stayed in the Cayo District is most appreciated. Thanks to Jacob and Kelly Marlin for providing a very comfortable base camp in the Bladen Nature Preserve. Both Sharon Matola, founder and Director of the Belize Zoo and Tropical Education Center, and Jan Meerman, Green Hills Butterfly Farm, shared important observations from their separate expeditions to remote parts of Belize.

Various conversations and collaborations with colleagues over the years have helped to develop this book. Particularly important in this respect were Bryant Capiz, Dr. Alan Jaslow, Jan Meerman, Dr. Gail Stratton, and Rick West. I learned about forest succession and tropical agricultural practices from my association with botanists Dr. Scott Franklin and Amy Webbeking, knowledge which was very useful during my study of tarantula habitats. The precise illustrations of emboli and spermathecae were the work of my mother, Norma Reichling. Dr. Thad Wasklewicz and Suzanne McMillian graciously assisted in the preparation of the distribution maps. I thank Bernhard Meck and Steve Bogardy for translating several German manuscripts. My friends from a television filming: Josh, Abe, Brian, Sara, and brave Holly, helped me collect wolf spiders and some of the tarantulas that were photographed for this book.

I borrowed numerous type specimens during my research and these loans were arranged through the courtesy of the following curators and their institutions: Dr. Manfred Grasshoff, Senckenberg Museum, Frankfurt; Dr. Laura Leibensperger, Museum of Comparative Zoology, Harvard University; Janet Margerison and Dr. Paul Hillyard, The Natural History Museum, London; Dr. Norm Platnick, American Museum of Natural History, New York; and Dr. Petra Sierwald, Field Museum of Natural History, Chicago.

I thank the Memphis Zoo, particularly Charles Wilson, Dr. Chuck Brady, Charles Beck, and Roger Knox, for the consistent support I've received. The Memphis Zoological Society provided funds for many of my field trips to Central America. Other trips which resulted in information presented in these pages were funded by Sigma XI and The American Arachnological Society.

# Chapter 1

# A Tarantula Hunter's Diary

The voice with the Aussie accent on the other end of the telephone line was full of assurance.

"Yeah mate, I think we can rustle up some tarantulas for you, our property is full of 'em!"

This was the news I'd been hoping to hear. For months, I'd been pondering how to transform my graduate school research from a plan that looked good on paper into a study that actually worked in the field. As every biologist knows, this process is one of the most difficult steps in field research. Unexpected kinks usually turn up during any experiment. In lab-based investigations, little glitches are expected but can often be overcome with a slight adjustment of the methodology. In fact, contending with such problems can be part of the learning process itself. Nature, however, cannot be "adjusted"; this role is reserved for the biologist. Field-based studies have to contend with a host of factors that never read the proposal, and the problems are often of a sort that conquering them doesn't contribute anything to the knowledge of the subject. Left unsolved, they simply leave the effort dead in its tracks. Bad weather, equipment failure, uncooperative local residents, illness due to everything from a bad bowl of stew to disease laden mosquitoes, are just a few of the ever-present threats lurking around the perimeter during a biological field study. And then of course, there are the animals themselves.

Pioneer tarantula ecologist William J. Baerg planned to study tarantulas in Jamaica. He had been awarded a fellowship to support the study. A sabbatical from his university had been secured. The family was packed up, and off they went. When they arrived in Jamaica, a quick scan of suitable-looking habitat failed to turn up any tarantulas. Concern began to mount when systematic searches were unsuccessful. Baerg discovered that the native tarantula of Jamaica had recently been extirpated by the introduced mongooses that have played havoc with so many West Indian animals. Having made a commitment to complete a yearlong stay in Jamaica to conduct research, Baerg marshaled that most important of character traits for a field biologist—adaptability—and set about studying the abundant black widow spiders instead.

The following narrative is taken from the pages of my field journal and is presented to provide the reader with a glimpse into the life of a field arachnologist. For every discovery about how animals make their living in the wild, there is a story involving people and places. The desire to follow a hunch about how an animal lives, and to follow that hunch into the forest requires the biologist to begin a difficult process. Yet it often becomes an obsession because the discovery of the truth about how the natural world works is such a prize that it takes only a little success to pay back all the effort expended during the search. Behind every scientific name in this text, and leading up to every description of behavior or ecology, there is a story. No truth about the most trivial facet of nature's mechanisms was learned without an interesting story attached.

I knew that for my study, the biggest challenge would be to locate a place where tarantulas were both abundant and accessible. My dissertation project was to find out the nature and extent of any care that female tarantulas might provide to their eggs or spiderlings. Once I determined that, I could compare it with data I'd collected in the laboratory under controlled conditions. Working late into the night at the University of Memphis, my wife Ann toiling diligently alongside me, I had been keeping busy raising almost 2000 baby tarantulas. Night after night, Ann and I would go up to my lab room, and begin the monstrously long process of individually feeding this mountain of tarantulas, one by one. Each spider resided in a small container—we discovered that urine specimen cups were ideal for this purpose—and we had to open each cup and drop in a cricket of appropriate size, squirt a little water into the cup to maintain adequate humidity, close the lid and return the container to the stack of finished cups. Each spider took only a few seconds to service, but repeating the process 2000 times took us several hours and was tedious beyond description. We got through this drudgery by reminding ourselves that it was necessary in order to get reliable data that would serve as a standard against which we could compare data collected in the field. We knew it would be worthwhile in the end, but I couldn't help but glance over at Ann occasionally and wonder how many other wives would give up several nights a week to endure this torture.

The lab study didn't disappoint us and our hard work was repaid with some interesting insights. We discovered that juvenile tarantulas are profoundly affected by how much food is available to them. The spiderlings surprised us with how little food they could survive on. We fed one group of spiderlings a near starvation diet consisting of one tiny cricket per week. The young tarantulas were amazingly hardy and tenacious but hardly grew at all. They survived, period. In startling contrast were their siblings that were living in a land of plenty. To this group we showered food three times per week, providing the largest crickets that the baby tarantulas could wrestle into submission. The spiderlings responded by showing us their seemingly insatiable appetite, and we saw that when given the opportunity, baby tarantulas will eat till they threaten to pop. In the popular lexicon of the southern states, the little gluttons looked "tight as ticks," and in fact they did resemble one of the parasites after a particularly large blood meal. These spiderlings grew and attained maturity very quickly, and after one year they dwarfed their deprived siblings by a factor of twenty. We discovered that the temperature at which the eggs had been incubated also influenced growth and speed of maturation, with a disparity of 5°C resulting in a measurable effect. Armed with the knowledge of what *could* affect spiderling growth, I wondered, were any of these factors being manipulated by female tarantulas to produce a particular quality of offspring? Discovering the answer to that question would provide insight into the forces of natural selection that had guided the evolution of tarantula physiology and behavior.

To determine what tarantulas were doing in the wild, it would be necessary to measure the temperature inside the burrows, some containing females tending cocoons and others which were not, and compare their behaviors. In order to do this in a scientifically meaningful way I would need to find tarantulas, and lots of them.

I considered working in the Arkansas Ozarks, which were only a 3-hour drive from Memphis. Tarantulas reach their eastern limit of U.S. distribution in the mountains of west-central Arkansas and Missouri. The big spiders are not found in the eastern sector of Arkansas, which is Mississippi Delta land and prone to flooding. In the Ozark Mountains and surrounding hill country, where sunny, rock-strewn slopes provide excellent habitat, tarantulas are abundant. However, to my disappointment they were not abundant enough to support the research I was contemplating. Ann and I spent several weekends driving all over western Arkansas, searching for a site where tarantula aggregations were suitably dense to be easily studied. We found plenty of medium-sized, brown *Aphonopelma* species during these trips, but we had to cover a lot of ground in the process. Unfortunately, the individual spiders were too widely scattered to be feasibly studied. To do so would require that I traverse miles per day to get my data. I would have to look elsewhere.

To measure comparative temperatures and to make observations on many individuals at roughly the same time of the day, I needed to find a place where many dozens of tarantulas were crowded together within a stone's throw of each other. I wondered if such a place existed. Everything looked great on paper, carefully outlined in the research prospectus I'd submitted to the graduate school, but putting the idea into action was turning out to be harder than I'd expected.

I was beginning to lose hope that a haven for tarantulas, such as I had imagined, actually existed anywhere outside the pages of my research prospectus. Then one day, I was chatting with a fellow tarantula enthusiast. I described my dilemma, to which he replied that he knew just the place I was looking for. He explained that on a recent vacation to Belize, tarantulas seemed to be everywhere, even though he wasn't specifically looking for them at the time (although I think he probably was because among people who are fond of tarantulas, the switch controlling the desire to seek them out is always on).

"Steve, I could have caught a bucketful in an hour if I'd wanted to," he said.

"But how much ground did you have to cover to find that many?" I asked, remembering Arkansas.

"It's absolutely incredible how dense they are!" he replied. "Just at a single site, around a pile of old boards, I must have seen a hundred."

What incredible news! I asked what kind they were.

"That's the only downside, they're all just redrumps, just common stuff."

I didn't really care what kind of tarantulas they were, unlike my friend, who was interested in finding rare species to add to his private breeding collection. Common Mexican redrump tarantulas, as *Brachypelma vagans* are called, would do just fine for my purposes. I wanted some wild landscape to serve as my laboratory, and I wanted the tarantulas to be as abundant as lab rats, and Belize was starting to sound like the place I was headed for.

"So, there's really only one species of tarantula in Belize?" I asked. I was a little surprised at the lack of variety in such tropical surroundings.

"According to the books, yeah, just *vagans*. But you know, I did find an all-black male tarantula squashed on the road running through a place called Maskall Village. I think it might be an undescribed species of *Brachypelma*." I knew this must have been an exciting discovery for him. The *Brachypelma* species are among the most sought after tarantulas in the pet trade, being beautiful, hardy, and often limited to small distributions—all adding up to an animal of considerable value to a private breeder.

"I'll kiss your feet if you bring me back a gravid female," he promised, although I already knew from previous experience collecting reptiles overseas for the zoo that a stipulation of most scientific collecting permits is that none of the animals collected, nor their progeny, would enter into commercial trade. No need to disappoint my friend with the bad news now, I thought. Besides, I probably won't find anything interesting down there anyway.

So Belize it would be. I had no idea that I'd just concluded one of the most pivotal conversations of my life. With Belize now in my sight, I'd taken the first step down a road which would lead to many years of exploration and adventure throughout the Yucatán Peninsula, and introduce me to lifelong friends.

Now my problem was to find a place to base myself while studying the spiders. I knew very little about Belize. I grew up reading the books of Ivan T. Sanderson, who made his living collecting animals during the 1930s and '40s and wrote several books chronicling his adventures. Sanderson's vivid descriptions of places like Belize, where he collected tarantulas, colored my naive expectations. From these stories I imagined a wild place teeming with strange creatures, as Sanderson had a tendency to emphasize the bizarre.

And now, following a lead from another friend who had been to Belize and given me the radiotelephone number of what he described as a very remote lodge, I listened as Colin Howells said the words I hadn't dared hope to hear.

"I reckon we've got more tarantulas around our property than anywhere in Belize," was his boast, and there was an intelligent tone in his voice that told me this wasn't an example of the sort of "tarantula hysteria" that often grips the uninformed when they see two or three spiders. For most people, seeing just a few tarantulas has the tendency to inspire tall tales.

Colin and I discussed my arrangements for coming down and working at the lodge, and in 2 months I was riding in a motor-powered skiff heading up the New River towards the lodge. By late afternoon the boat pulled up to the dock jutting out from a steep sloping bank decorated with an array of thatched buildings of various shapes and sizes. The structures were built in a clearing in the midst of forest, and the scraped appearance of the ground and numerous tree stumps with the cut ends still white indicated that the site had only recently been claimed and opened.

Mark Howells, Colin's son, was at the dock to welcome me. A brick wall of a man at 6 feet 4 inches tall and shaped by a life growing up in the Australian outback, my first impression was that here was a tough man who could handle any crisis with ease. As I began to know him better I learned that my first impression was correct and that it was the universal assessment, as proven by the respect Mark was given by everyone in Belize. We walked up the hill together on the way to the cabaña where I'd be staying, and the conversation turned to tarantulas.

"Mark, your dad tells me there's a lot of tarantulas here on the grounds. Where's the best place to start looking?"

"Steve, I think you've already found yourself a couple," Mark said, pointing to my feet. I looked down and there within a square foot of each other, right beside my feet, where two huge burrows, their openings protected by a sheet of silk. Unbelievable!

"Is this typical for around here?" I asked, affecting a restrained monotone. I was trying not to appear too strange to my new acquaintance by acting as giddy as I was feeling.

"The ground is like Swiss cheese there's so many holes," Mark grinned, "but you're the first person who ever came asking about 'em!"

I scanned the ground as we finished our hike up the steep hill and casually counted another dozen burrows in the fading late afternoon light. At the main building and dining room, I was greeted by Mark's wife Monique, who shook my hand while balancing a young black howler monkey on one shoulder and a yellow-fronted amazon parrot on the other. Monique had gained a local reputation of having a gift for raising orphaned and injured animals, and these two, Chinko the monkey and Poppy the parrot, were just a few of her children. To Monique's credit, every animal she reared was done so with the intention of releasing it back into the forest after a gradual transition and training period. Knowing what was truly best for the animals, she never gave in to the tug of her heartstrings by turning them into permanent pets.

I heard a familiar voice and turned to see Colin Howells come around from behind the building.

"Steve, you won't believe this, but we've been getting ready for your arrival. Ever since talking to you I've had the guys that work around here bring me every tarantula they could find." This wasn't exactly what I had in mind, because I wanted to study the spiders while still in their burrows, but I did intend to do some experiments that would involve captured animals so these might come in handy for that, I thought.

"Where are they?" I asked.

"Right where they'd keep for you, Steve." I was puzzled.

"What do you mean, 'where they'll keep'?" I asked him.

"Well, 200 dead tarantulas are gonna turn bloody awful in a few days if we don't ice 'em down, I reckon, so we've been keeping them in the freezer for you. There's a right big pile of them in there," was his shocking reply. Dead? I never said anything about wanting them dead! This will ruin everything. Tarantulas were obviously very common here, but if they've killed even close to 200 adults, it would have a huge impact on the local population, not to mention my research. My heart was beginning to sink, and I was feeling sick. A grin Mark was struggling to suppress caught my eye.

"What's going on?" I asked, not having the slightest idea.

Colin was now smiling broadly. "Aw, I'm just kidding, we wouldn't harm a hair on a tarantula's body around here. Had you going though, eh?"

"Steve's just gotten here and you're already getting cheeky with him!" scolded Monique. "Let him get to know us first before you start with that." But I already was getting to know Colin and Mark, because I soon learned that they were both masters of the straight-faced put-on. Fortunately, from that point on I was usually an inside observer as they played pranks on others, and on their favorite targets of all, each other.

That evening I ate a hurried dinner and began getting acquainted with my hosts. Seated around the table were Colin, Mark, Monique, Colin's wife Ellen, and two guides who worked at the lodge, José and Carlos. My mind was only partially following the conversation, because I was so excited thinking about what I'd find when I headed out into the surrounding forest once night had fallen.

"I guess you'll want to turn in early tonight so you can go looking for tarantulas tomorrow," Monique suggested, and her voice, directed at me, pulled me back from thoughts about the forest.

"Actually I plan on doing some looking around this evening, so I probably ought to be heading out soon. I saw some thick forest on the bank north of here when I was coming in on the boat, so I think I'll check that out."

"I don't recommend that," Colin warned, "there are surprises out in the bush at night you don't want to run into."

"Dad's right," Mark agreed, "you don't want to get surprised out there."

Everybody around the dinner table had become interested in contributing to this conversation and looked eager to pipe in. "Surprises, I'll say," Ellen said emphatically, "like the time I took my afternoon walk to the ruins and a jaguar followed me. I caught something out of the corner of my eye, a little patch of orange." Everybody was listening intently, but I knew they'd heard this story before. They were interested in my reaction to it. "It just followed me," Ellen continued, "I wanted to run but kept telling myself to be calm. He shadowed me all the way back to the lodge. Every time I stopped, he stopped, and when I walked, he kept pace with me and was right beside me so close I could hear him panting."

"So you stayed calm and everything went all right?" I asked, but I was trying to end the story with a moral.

"Right, but if that'd been at night, the cat might have been less interested in playing and more interested in eating," Colin said firmly.

"And don't forget about snakes," Carlos warned.

"I don't mind snakes," I said, "In fact, I'm inordinately fond of them," I added, trying to sound cute.

"Sure, but at night they're all over the trails and are almost impossible to see. You'll be stepping on them. The yellow-jawed tommygoffs get up to 7 feet and are deadly," Carlos countered my dismissal of the danger of snakes. "Very aggressive," and he nodded knowingly to Colin.

José hadn't said anything all evening but everyone now turned to him, expecting him to speak, as a stage had apparently been set. "The Caretaker of the Forest doesn't want you there at night," he said.

"The Caretaker of the Forest? What's that?" I asked.

"The Maya say the Caretaker looks after the trees and animals. He gets mad if you hurt anything there, and will punish you if you do," José replied.

"I don't go in to hurt anything. I just want to look at tarantulas," I said.

"He doesn't want people in there after dark. That's time for things you're not supposed to see." José's face was stiff and wooden. I could see this was not a put-on.

"There are surprises out there, mate," Mark spoke up as José leaned back in his chair, signaling he would say no more, "Just be careful."

I departed for my first foray into the forest with instructions from Mark on how to follow a trail to a milpa tended by Enrique, who lived in nearby Indian Church Village. Carlos and José both agreed that Enrique's milpa was full of tarantulas. I found it easy enough, about a mile from the lodge. It was about the size of a large backyard, a gash chopped out of the surrounding tangle of trumpet trees. It was late August, so planting had taken place 4 months earlier and now the ground had become overgrown with shin-high grass, providing ideal conditions for chiggers as I discovered too late.

The tarantulas lived up to the boasts. They were everywhere. Within an hour I had located 60 burrows and marked each with a bright red flag. I shone my flashlight down each hole to see what species they contained. Inside every one were two pairs of velvety black forelegs, a black carapace, and a few red hairs peeking out from behind the cephalothorax. Just as my friend had told me, Mexican redrumps, and plenty of them, but nothing else. I peered down into another burrow, expecting the usual, but something different was there. It was a tarantula, but brown with no black or a trace of red. I utilized my newly acquired skill at tarantula fishing, having already practiced on a dozen Mexican redrumps, and out came a beautiful creature. Clothed in short, silky hair, not woolly like a redrump, I held a female *Crassicrus lamanai*. This was a species that had been an occasional feature on dealers' price lists for years as the "cinnamon tarantula" but which had never been formally described, named, or classified. At the time they were usually labeled "*Citharacanthus* sp." Knowing I had something interesting, but over a year away from figuring out exactly what, I gently placed the big spider in a deli cup and continued searching.

I didn't have to wait long for the next discovery. At the entrance to a burrow I saw a male tarantula, and he was tapping out a rhythm on the rim in an effort to lure the female out. The spider was solid black. Incredible, I thought. This is the black *Brachypelma* my friend had talked about so longingly. I also remembered that there was no *Brachypelma* species known which was solid black, the closest being specimens bred in Europe from stock supposedly collected somewhere in the Yucatán. These spiders, possibly *B. epicureanum,* have a sparse amount of long red setae on an otherwise black body. The tarantula at my feet had not a trace of red. What luck! I'd be able to collect the female too, which would be essential for ruling out any known species and forming the basis of an informative new species description. I adjusted the direction of my flashlight's beam slightly to illuminate the interior of the burrow. Sitting just a few inches down and drumming back eagerly to the strange black male was a female cinnamon tarantula! I couldn't believe my eyes, this male was so unlike the female in appearance. No wonder my friend hadn't made the connection. He knew of the cinnamon tarantula only from the females. The adult males had never been seen. I was a little disappointed that the black *Brachypelma* had turned out to be a myth, but reminded myself that I had found a new species (later turning out to be a new genus as well), and both sexes, during my first hour of looking in Belize.

One night, about a month later, I was out wandering the grounds of the lodge. A light rain had begun, but massive lightning discharges flashing to the north warned of a big storm on the way. During the rainy season these storms come almost every night, and although the lightning and thunder are usually not as powerful as storms in the United States, the unbelievably heavy downpours are so intense that the sound of the water roaring against the thatched cabaña roof is more frightening. Although a sudden clap of thunder is startling, it seems unfocused and oblivious to your presence, but these titanic downbursts seem more determined, a more personal threat. I didn't want to be too far when it really opened up, so I strolled close to the lodge. I had been walking, not looking for anything in particular, and had paused to watch the bursts of lightning in the sky, now close enough to brighten my arms with each flash. I looked down, and panned my light across the ground, rather aimlessly. The edge of the beam caught a familiar shape silhouetted on a white limestone boulder. After weeks spent looking for tarantulas and finding hundreds, a black spot with eight lines radiating outward matched a search image that was seared in my mind, and the sight sent a jolt of adrenaline through me. It was a tarantula. I focused my light down on the spider. It was a Mexican redrump, a male by the look of him, thin and spindly and smaller than any mature Mexican redrump I'd seen. I had no reason to collect him, and began leaning into my first step, preparing to move on. But there was something strange about this spider I couldn't place my finger on. I leaned down and took a closer look. I confirmed the presence of the bright red abdomen and darker carapace and legs, although this one did seem a little washed-out. I almost walked away a second time, and to this day I don't know why, but I stopped, leaned over, plucked the spider off the rock, and dropped him in a cloth bag I had looped over my belt in case I found an interesting snake. After another half hour, by which time the storm was almost to the lodge, I went back inside my cabaña and tumbled the spider out on my bed to have a better look. This was no redrump. In the bright light of the lamp beside my bed, I could see a gorgeous golden iridescence playing over the carapace and legs. The abdomen was orange, not red, and of a fiery intensity. I picked up the spider and brought it close to the lamp, turned it over, and saw that it was indeed a mature male, but that there were no tibial spurs. All male tarantulas in Belize had tibial spurs, I told myself. I had brought down no manuscripts or keys with me on this trip, so it was not until I returned home that I found that this was an undescribed species—Belize's most beautiful and probably its rarest—Gutzke's tarantula, *Metriopelma gutzkei*.

To be a successful observer of tarantulas, a person has to become comfortable roaming around in the forest, at night and alone. Of course, there is never a shortage of interesting company, but it's the nonhuman kind and this is unnerving to many people. I have to confess that when I first began my forays into the Belizean forest I sometimes got the strong sense that some presence was following right behind me. I tried not to think about what José had said. I would ignore the feeling as long as possible but it steadily grew until I had to stop and turn around, panning my flashlight out into the dark like a fool. There was never anything there. After some weeks went by, the sensation disappeared and I was left with only a very pleasant awareness of being part of a community, just a member of a bustling place where creatures of all types were carefully but contentedly going about the business of their lives, each with important chores to accomplish. The gibnut crunching on cohune nuts, the clumsy armadillo plowing through a tangle of lianas, kinkajous, locally known as nightwalkers, carefully plying their way through the canopy and plucking sweet figs, even the venomous tommygoff coiled alongside the trail I was walking on was simply a member of this community that I became a part of each night. No harm ever came to me in the forest. I made my way carefully, trying not to frighten the other creatures, for surely to them I was the equivalent of the monster that most of us fear in the back of our minds when we walk alone in the woods at night.

Being comfortable while out at night freed me to do it often and for long periods, and most importantly alone, because every Belizean animal except the jaguar fears some predator, and their senses at night are at their most finely tuned. Only a solitary person, walking quietly in soft shoes, has much chance of seeing most of the animals before they flee.

But what about the flashlight, you may be asking. Doesn't that alert the animals to your approach long before you can see them? I carry a flashlight but normally don't use it until I see a shadow or hear a noise close by. Particularly when there is a good moon out, but even at other times after the eyes have acclimated to the dark, you can usually see indistinct dark objects as they move across the path in front of you, or key into the sound of soft footsteps in the surrounding brush, without the aid and hindrance of an artificial light. I think our ears and bodies in general are more perceptive to nonvisual stimuli than we commonly require them to be, and we rely on our eyes and let our other senses atrophy and grow dull.

Without a willingness to become as nocturnal as a tarantula, most of their behaviors will always be out of sight. Had I not been strolling down a jeep track through the Lamanai Reserve one evening, I would never have witnessed one of the most amazing sights I've ever seen while studying tarantulas. I was walking the road because the white limestone marl which was laid bare by the passage of vehicles made it easy to spot small creatures lying there. I was scanning the road carefully as I walked, hoping to find snakes since this was a particularly hot and sticky night, when about 2 meters in front of me I saw what I took to be a small, slender snake, probably a coral snake, or so I thought. I pulled my flashlight out of my pocket and shone it on the thin, dark line that was undulating slowly across the track. The image caught in the light had me rubbing my eyes, exclaiming words of surprise to no living thing that could understand me. It was a procession of baby Mexican redrumps that had just left the maternal burrow. They were heading out into the unknown, together, to begin their new lives. Up until that moment this behavior was unknown for any spider, yet they'd been doing this for thousands of years, under cover of night.

I walked back to the lodge, arriving at midnight and found Mark and Monique still up, waiting for the staff to finish closing the dining room. Mark offered me a beer and asked if I'd seen anything interesting.

"Well Mark," I said, "let me tell you about a surprise!"

# Chapter 2

# Physiography and History

### Regional Overview

Naturalists writing about Belize often comment on the disparity between the country's landscape diversity in comparison to its small area. Both these traits make Belize an interesting place to study tarantulas. The small distances between sites, further reduced by a good road system, make virtually all corners of Belize accessible. Only the interior areas of the highlands are remote. In other Central American nations there are extensive areas of lowland (e.g., Panama's Darién) or highland (e.g., the Cordillera Isabella in Nicaragua) that are very difficult to reach, and which make a thorough biological survey a goal whose attainment is many years away. In Belize, within a single day it is possible to visit habitats that are hidden away in other countries. Belize lacks the dry tropical forests and montane communities of Costa Rica's Guanacaste and Monteverde parks, but examples of most other habitats relevant to tarantulas are present.

Belize also has a varied geological history which enhances its arachnological appeal. Most tarantulas are creatures of the soil, so variations in coarseness, compaction, moisture, drainage, and slope will influence species composition. Ancient granitic soils in the Mountain Pine Ridge are dry and hard, and this region is depauperate of species. Moving further inland, the substrate gives way to soft, moist soils on the Vaca Plateau that were derived from Cretaceous shales and mudstones. Limestone-based marls are mixed with sands around the perimeter of the highlands, creating loose and well-drained ground which is conducive to fossorial spiders. The northern and northwestern sectors are a Tertiary limestone shelf with deep sediments of quartz and silica sands originating from Quaternary alluvial deposition, and at least one species is essentially restricted to this region. Along the coastal plain is an area of recent sedimentation consisting of mud and clays, and where suitable tarantula habitats are disjunct owing to an abundance of wetlands and saturated soils.

Abrupt transitions in topography are widespread. A drive up the Chiquibul Road from Georgeville to Bald Hills, a distance of under 25 km (15.5 mi) climbs over 800 m (2624 ft). As elevation increases, so does the character of the terrain and with it, a corresponding change in the tarantula fauna. Stone outcrops and granitic soils that appear as the Mountain Pine Ridge is reached delineate a zone of transition, where the two genera which dominate the lowlands, *Brachypelma* and *Crassicrus,* give way to the genus *Citharacanthus*. Although *Brachypelma* reappears further into the interior of the Maya Mountains, where the soils again become rich and moist, it is never again as abundant as it is below the 200-m line.

Two spiders, the Livingston tarantula, *Citharacanthus livingstoni,* and the Costa Rican orangemouth tarantula, *Psalmopoeus reduncus,* reach the northern limit of their distribution in southern Belize. These incursions are likely due to Belize's pronounced north-to-south rainfall gradient. The Shipstern Peninsula, at the extreme northeastern corner, is the driest region, receiving an average of 135 cm (53 in) of rainfall per year, most of it during a distinct wet season (June–December). During the dry season the ground becomes hard-baked and blanketed with leaves cast off by the deciduous scrub. Under these conditions, only deep-burrowing tarantulas are able to survive and arboreal species are excluded. Points farther south receive progressively more rain, and the line between wet and dry seasons becomes blurred. However, these variations are too slight to impact the biogeography of tarantulas, resulting in the majority of Belize resembling the rest of the Yucatán region in terms of species composition. In the extreme southern section of the country, however, the abundant and nearly aseasonal rainfall causes the forests to bear more resemblance to the tropical lowlands of the rest of Central America than the comparatively xeric and less lush Yucatán. Annual rainfall amounts near the town of Punta Gorda, in Toledo District, may exceed 453 cm (178 in). Consequently, the tarantula fauna is somewhat different, comprised of species not found elsewhere in Belize.

The result of these contrasting factors is a tarantula fauna of surprising diversity for such a small geographic area. The northern lowlands, part of the Yucatán Peninsula both geologically and biologically, harbor a distinct assemblage of species. The Maya Mountains and highlands to the lee are an eastern extension emanating from Guatemala and have their own fauna. The moist tropical forest formations of Honduras make a small inroad into southern Belize, contributing two species.

## Historical Background

The first tarantula discovered in the Yucatán Peninsula is also the most common and widespread species, *Brachypelma vagans,* described and named by the Austrian arachnologist Anton Ausserer in 1875. Ausserer also described a second species from the region, *Ischnocolus sericeus,* on the basis of a single juvenile tarantula. Tarantula taxonomist Andrew Smith examined this spider in London's Natural History Museum and found that its immaturity and lack of any distinctive characteristics made species identification impossible. Smith therefore proposed that the name *sericeus* be disregarded as a valid species, a recommendation which I have followed.

The year 1897 was an important one for our knowledge of Central American tarantulas, for it was then that British arachnologist Frederick Octavius Pickard-Cambridge produced an initial survey of the tarantulas of the region and described a number of new species. Like most taxonomists of the time, Pickard-Cambridge was not a field biologist. He classified preserved specimens which had been collected by explorers working overseas and who were sending material back to Europe, material that eventually made its way into The Natural History Museum where Cambridge was curator.

## Physiography and History

The study of Yucatán tarantulas has, since its inception, been closely associated with the excavation of the ruins of the Maya civilization. During the late 1800s and early 1900s, the lowland forests in the Guatemalan State of Petén and most of present day Belize were virtually inaccessible to biologists. Among the few outsiders with scientific interests to venture into this region were archaeologists. Not only did these explorers penetrate to remote sites, but because their work involved careful digging and displacement of stones, they inevitably encountered tarantulas that other early adventurers overlooked. It is not surprising that among Pickard-Cambridge's eight Central American species descriptions, the only one from the Yucatán was found near the ruins of the Maya city of Tikal. By the close of the 19th century two tarantula species were known to occur in the Yucatán Peninsula, the other species still hidden and unsuspected.

No further progress was made for a quarter century. In 1925 Ralph Vary Chamberlin, an invertebrate taxonomist and curator at the Museum of Comparative Zoology at Harvard, described two new species that had been collected among the ruins of Chichén Itza in the state of Yucatán, Mexico. These species, *Eurypelma stoica* and *Brachypelma epicureanum,* apparently do not occur as far south as Belize.

For over a century, the only species of tarantula known to occur in Belize was the Mexican redrump, as this area had never been studied by anyone interested in these spiders. The distributions of Pickard-Cambridge's species were unknown, and neither of Chamberlin's taxa occur in the country. Then in 1996, Gunther Witt described the Maya tarantula, *Psalmopoeus maya,* from a cave in the Cayo District, the first indication of the diversity that still awaited discovery. In 1994 I began exploring Belize in search of tarantulas and to study their natural history and distribution and, in the tradition established by earlier biologists, the first place where I studied was the archaeological site at Lamanai.

# Chapter 3

# The Tarantula in Context

### Basic Spider Classification

Spiders, Order Araneae, along with the crustaceans, are the only extant members of a group (probably polyphyletic) known as "chelicerates." They are distinguished from insects by the morphology of their mouthparts. Two broad categories, or suborders, of spiders are recognized: the Opisthothelae and the Mesothelae. The mesothelids are a primitive, poorly known group with a few species living in eastern Asia and Japan. Two striking mesothelid characters that emphasize their primitive form are abdominal plates and segmentation, which is the ancestral condition for spiders. The Opisthothelae is subdivided into two smaller groups, or infraorders, the Mygalomorphae and the Araneomorphae. Araneomorphs are the "typical" spiders that represent the great majority of the over 36,000 named species of spiders in the world. Typical spiders include such familiar kinds as garden spiders (family Araneidae), and the widows and cellar spiders (family Theridiidae). Wolf spiders (family Lycosidae) (Fig. 3.1) and wandering spiders (family Ctenidae) (Fig. 3.2) are common and conspicuous araneomorphs in Belize, and, due to their large size and hirsuteness, are often mistaken for tarantulas.

Figure 3.1 Fossorial wolf spider, *Hogna carolinensis*. (Photo by Gail E. Stratton)

Figure 3.2 Wandering spider, *Cupiennius* sp.

The infraorder Mygalomorphae includes the tarantulas as well as trapdoor spiders, funnelweb spiders, and purseweb spiders, for a total of just over 2400 species. While not as prehistoric in appearance as the mesothelids, mygalomorphs have retained many primitive characters and can be distinguished from typical spiders by these traits. The fangs of typical spiders move perpendicular to the long axis of the body, in scissorlike fashion (much as your thumb and forefinger move toward each other when you pinch something). Mygalomorph fangs move parallel to each other, with the long axis of the body (as when you retract the first and second finger on one hand in unison). When retracted and not in use, tarantula fangs lie side by side in nearly parallel alignment, rather than with the tips turned toward each other in pincherlike fashion (Fig. 3.3). Araneomorph ventilation is through a tracheal system, as in insects, with one external ventral spiracle, and one pair of book lungs (a stack of very thin tissue through which gas exchange can occur, reminiscent in form of the pages of a book). Mygalomorphs display the primitive condition: one or two pairs of book lungs but no trachea or spiracle. Typical spiders have one pair of spinnerets (fingerlike structures used to manipulate the silk as it's secreted by the silk glands) while mygalomorphs have a tiny anterior pair and a much larger, functional posterior pair which are clearly visible (the primitive condition is believed to be more than two pairs).

## Fossil History

The earliest fossil spiders are assignable to the Mesothelae and Mygalomorphae. The oldest known spider is the mesothelid *Atterocopus fimbriunguis* from

# The Tarantula in Context

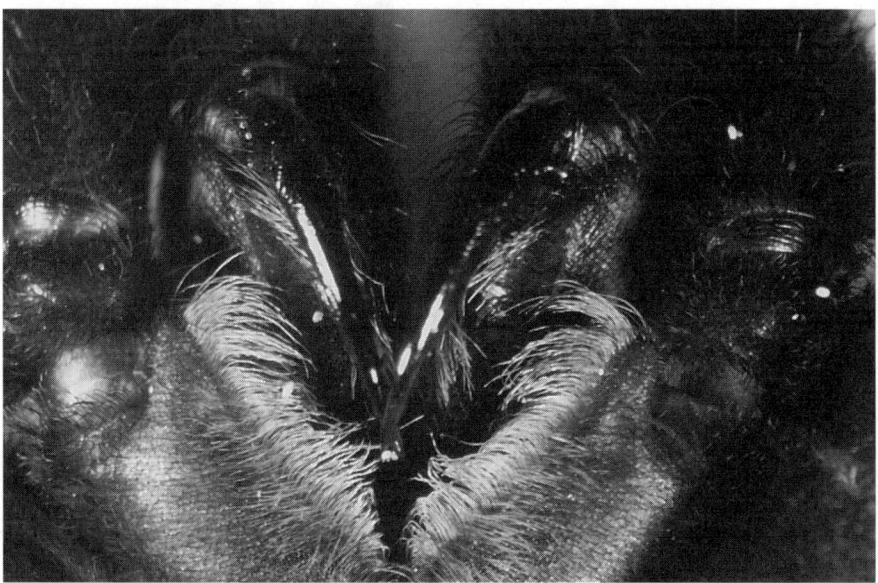

Figure 3.3 Tarantula fangs.

the middle Devonian Period (408–360 mya). The oldest mygalomorph fossil remains are that of *Rosamygale* from the lower Triassic Period (248–231 mya), which exhibited primitive characters such as abdominal plates and six pairs of spinnerets. The oldest tarantula remains have been recovered from Dominican amber beds, and date back to the Miocene Epoch (24–5 mya), the oldest named species being *Ischnocolinopsis acutus*. Although approximately 98 percent of spiders living today are araneomorphs, the mygalomorphs are believed to have been the dominant group at the time of *Rosamygale*. From the spider's standpoint, the Mesozoic Era could be called the "Age of Mygalomorphs." Today we know of over 850 species of tarantulas, and certainly many species remain undiscovered. Nonetheless, this group of large, primitive spiders is probably less specious than it once was. The evolutionary success and radiation of modern plants, with the subsequent increase of flying insects, prompted the dominance of the araneomorph spiders because the new prey required a new method of capture—the web.

## Types of Tarantulas

The designation of a spider as a "tarantula" is somewhat subjective. Most often, this designation is used for one family of spiders, but there are other types of mygalomorphs that fit the general physical characteristics which define this group. The mygalomorphs are composed of fifteen families, including such familiar kinds as the trapdoor spiders (Ctenizidae) and the deadly Sydney funnelweb spider, *Atrax robustus* (Dipluridae), from Australia. Several of these families, including both mentioned above, have representatives in Belize. However, it is two other mygalomorph families, Theraphosidae and Barychelidae, that warrant the name "tarantula" and are the focus of this book.

The Theraphosidae family is subdivided into a contentious group of ever-changing subfamilies, reflecting the embryonic state of systematics for spiders in general and this group in particular. Several subfamily arrangements have been proposed. For the most part, I follow Robert J. Raven's 1985 revision, in which four Old World and three New World subfamilies were defined, although with a few departures from this scheme as a result of more recent studies. Three theraphosid subfamilies are relevant to the scope of this book. The large Ischnocolinae subfamily has fluctuated in scope, once including both Old and New World species, then restricted to Old World, and recently returned back to include tarantulas from both hemispheres (according to analysis by spider systematist Jan-Peter Rudloff). One Ischnocolinae genus, *Acanthopelma,* has its sole species occurring in Guatemala, and possibly in Belize. The predominantly Old World Selenocosmiinae includes two Belizean species of the genus *Psalmopoeus*. The largest subfamily in the New World is the Theraphosinae, which might be called the "typical tarantulas," as these are the large, hairy, ground-dwelling spiders that enter most people's minds when they think about tarantulas. Most of the species covered in this book are theraphosines.

The second family of tarantula spiders treated here is Barychelidae. Theraphosids and barychelids appear to have descended from a recent common ancestor, and share many similarities of form and function. Barychelids are never as large as what people typically expect for a tarantula, but in other ways they are quite similar. Belizean barychelids live in silk-lined burrows, are compact and stocky, and walk with the slow, graceful, faintly ominous gait characteristic of other tarantulas. One of the reasons they are not always considered when speaking of tarantulas is that the taxonomic designation is rather arcane outside arachnological circles, and most people encountering a barychelid would simply assume they were small theraphosids, i.e., a tarantula. To recognize the distinction between barychelids and theraphosids, I refer to barychelids as pygmy tarantulas. Two pygmy tarantulas occur in Belize, both members of the sufamily Trichopelmatinae.

# Chapter 4

# Why Care About Tarantulas?

Tarantulas have a huge public relations problem. Most people accept the myth that tarantulas are aggressive and sinister creatures lurking the jungle and looking for any opportunity to pounce on an unsuspecting human victim to inflict a deadly bite (Fig. 4.1). The truth about tarantulas couldn't be more different, but it's very hard to get most people to see this. In general, attitudes toward other persecuted animals like snakes and large predators seem to be improving, but tarantulas still are mostly loathed. This is a shame because tarantulas are actually very shy and beautiful animals. Although it may be difficult to impart an appreciation of tarantulas to most people, it should be possible to teach tolerance through the development of an environmental ethic (Fig. 4.2).

In general terms, an environmental ethic implies that individuals feel a connection with nature and all living things, and conduct their behavior accordingly. For biologists this entails treating the organisms they come in contact with in a respectful and gentle way, not only in terms of direct individual contact (careful handling, meticulous husbandry, euthanasia and invasive procedures done only when well justified and in the most humane way possible), but also indirectly in

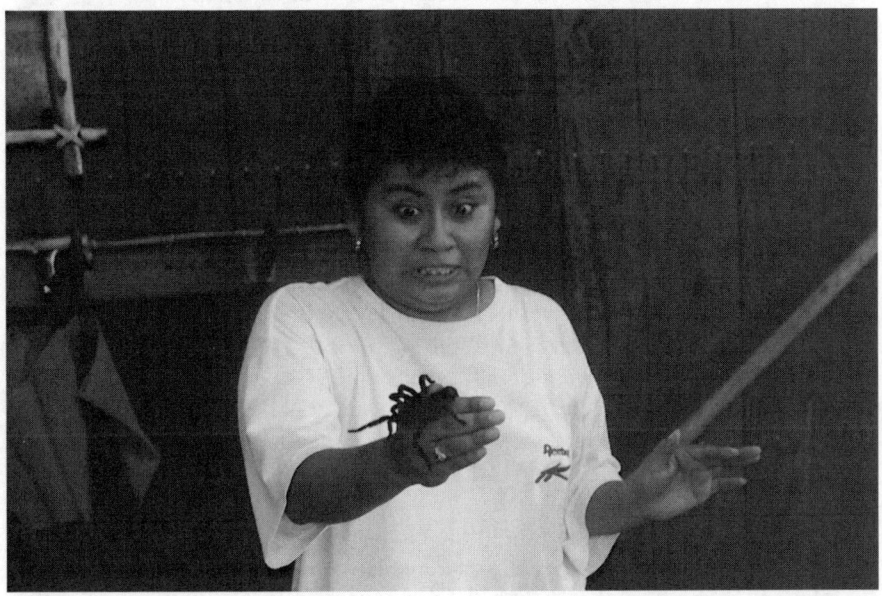

Figure 4.1 Typical response to tarantulas based on misconceptions.

Figure 4.2 An attitude based on understanding.

the way they treat the habitat where their work is being conducted. For example, no biologist with an environmental ethic, wishing to study fish, would electroshock a pond when seining and release would yield sufficient data.

Two individuals, each writing in a way that pushed conventional limits of thought in his time, illustrate the development of the environmental ethic in biology. Although neither man had tarantulas on his mind, their philosophies are very relevant to anyone trying to promote tolerance for misunderstood organisms. Henry David Thoreau knew what an environmental ethic was, and *Walden* is the strongest argument for embracing an environmental ethic ever written. Thoreau recognized the intrinsic value of every life form, no matter how small or inscrutable to human eyes it appeared to be. This is an important quality to have when considering misunderstood animals like tarantulas. The beauty of spiders is easy to overlook, but enriched with a reverence of all lifeforms, as Thoreau taught, it becomes a joy to pause and admire the smallest of creatures. Most biologists possess this insight even as children, perhaps are even born with it, and it is this inherent ability to appreciate the tiny creeping and crawling entities that compels them to make careers based on observing animals. In order for the majority of people to care about creatures like tarantulas however, the environmental ethic must be taught.

It was Aldo Leopold who brought the environmental ethic to the forefront of public consciousness, and who really attempted to convince society that such a philosophy was its responsibility. Leopold's *Sand County Almanac* was a plea for conservation thinking and is required reading for the student of tarantulas. Leopold used the term "land ethic," repeatedly calling for a sense of land stewardship, and assigning worth to the natural world and all its inhabitants. Leopold's description of his disdain for myopic "academicians" who professed to understand biology but held no love for its subjects is something every naturalist needs to keep in mind always. Leopold felt that to be a good steward one must love the entity.

There should be no taxonomic chauvinism associated with an environmental ethic—if it's right, it's always right. Everyone who claims to care about nature and wildlife must love all of nature. Anyone who wants to learn about tarantulas with any prospect for gaining a true understanding of them needs to internalize this message.

Belizean tarantulas are not aggressive animals. Often, they can be coaxed from their burrows and handled immediately. It is true that very few freshly captured wild animals could be handled without the likelihood of the person being bitten or scratched. Yet in spite of this peaceful nature, very few animals in the tropical forest are as widely feared as the tarantula. So many false and negative stories have been told for so long that they have become the conventional wisdom and are accepted without question. Tarantulas are a staple in adventure stories set in the tropics, and their role is always sinister. The truth about tarantulas is that they are shy, tolerant of mistreatment, extremely beautiful in form and color, and graceful in movement.

People want to know, why should they care about tarantulas? A woman I met in Belize one summer—an eco-tourist interested in birds—asked me what my tarantula research would mean to her. Why should she care? I avoided the cliché retort that perhaps some pharmaceutical wonder lay hidden in the physiology of these spiders, that maybe someday we might learn to regenerate amputated limbs by studying the way tarantulas can do this. As far as I can tell, the quality of human life would not be affected in any way other than aesthetic by the loss of any tarantula species. Leopold said, what is valued can then be devalued, meaning that when we place a monetary or human utilitarian value on a creature, we cheapen it by making the organism an obstacle to getting what we really cherish after all— something of use to us. A frog in South America which has very poisonous skin secretions (one from the family of frogs which are used by some native people to poison their blowgun darts) proved to be the source of a pain killer as powerful as opiates but without unwanted systemic effects. However, within a few years the compound had been synthesized. From obscure, to precious, to obsolete in just a few years, and now this frog is no longer needed by mankind, nor the habitat it lives in. This is not a solution to encouraging reverence of wildlife.

I answered the woman by telling her that I studied tarantulas because I liked them—how they looked, moved, lived, where they lived—everything about them. Then I showed her a cinnamon tarantula I had just caught, and told her about how the female ardently tends her cocoon for 2 months until the babies emerge. I described how she would defend her burrow against predators as best she could, and how she was especially adamant in doing this while she had an eggsac. I explained to the woman that this spider was at least 10 years old, and that others like her might live in the forest for another 10 years more, surviving, just trying to live and wanting to be left alone. I demonstrated how gentle the spider was, how completely unlike the popular conception this shy creature was. Soon she was enjoying watching the spider and commented perceptively on how graceful its movements were. The next day, she called me away from my study site to show me something she had found, a tarantula exuvium. She was fascinated with it and I doubt that she ever stomped on another spider again. That kind of appreciation lasts forever and enriches the life of anyone who ventures out into the woods or jungles, or backyard. It is the essential foundation upon which, hopefully, another personal environmental ethic is built.

# Chapter 5

# Why Worry About Tarantulas?

At present, just over 850 species of tarantulas are known worldwide. Several new species are described every year and it is certain that, as with all tropical invertebrates, there are many species still awaiting discovery. Unfortunately, with destruction of the tropical forests proceeding at a rate of 15 million ha (~37 million acres) per year, many species could become extinct before they are found.

Among tropical organisms, tarantulas possess several vulnerabilities which put the survival of many species at particular risk. Beauty, variety, and rarity are traits which attract collectors of jewels, stamps, art or virtually anything else that people have found desirable to possess and covet. Tarantulas have these qualities and it is quite easy to think of them as living jewels. The value placed on collectibles increases with demand and desirability, and so it is true with tarantulas that as laws have been passed to protect them, the supply of legal specimens diminishes and it becomes an elite status symbol among some collectors to possess rare and "unobtainable" species.

During the 1960s, the Mexican redknee tarantula, *Brachypelma smithi,* a beautiful spider from Mexico's Pacific coastal region, became a popular pet in the United States and Europe. To supply the demand, this species was exported by the thousands every year. Although a common spider, redknee tarantulas were collected in such great numbers that the Mexican government and conservationists elsewhere grew concerned that some local populations might be severely depleted. On 5 June 1986, in an effort to regulate commercial trade and protect the species, the Mexican redknee tarantula was listed by the Convention on International Trade in Endangered Species of Wild Fauna and Flora (CITES) as an Appendix II species.

CITES is an international agreement among participating countries (139 as of 1999) to cooperate in regulating commercial trade of each other's endangered plants and animals. Prior to CITES, once a plant or animal had been illegally taken out of a country, it was unlikely that it would be stopped as it entered another country since it was difficult for wildlife agents to know what the laws regarding wildlife export were in every other nation. In the United States for example, as long as a particular species wasn't protected by the U.S. Endangered Species Act and did not come under the jurisdiction of the U.S. Department of Agriculture as a potential plant pest (in both cases tarantulas were not affected) the wildlife inspectors often had little choice but to clear the shipment, even if it had been illegally collected and smuggled out of the country of origin. In 1981 an amended version of the 1900 Lacey Act was passed and mandated that the U.S. Fish and Wildlife Service (USFWS) would enforce the wildlife laws of other nations when species were illegally brought into the country. However, it was still problematic to know which species were protected by which countries and to what degree. Since the advent of CITES, every country has the opportunity to list native species that it believes may be endangered or threatened by commercial trade. Species are categorized in one of three levels of protection: Appendix I, II, and III, with Appendix I being the level with the most narrow conditions under which trade is permitted. The Appendix II listing of *B. smithi* means that no specimens, either alive or dead, may be exported without a CITES export permit issued by the CITES Management Authority of the country of origin (specimens captive-bred and living in countries other than Mexico are also covered by CITES rules), and wildlife authorities worldwide have agreed to enforce this regulation. Each country where the listed species naturally occurs makes the determination regarding what number of specimens will be permitted to leave the country, and often yearly quotas are established. In the case of the redknee tarantula, Mexican wildlife officials have not issued CITES permits for any specimens slated for commerce since the species was listed.

Since there have been no legal exportations of wild-caught Mexican redknee tarantulas for the purpose of supplying the demand for pets since 1986, commercial interest shifted to other attractive species, particularly other *Brachypelma* spp. As a result, in 1994 all species of *Brachypelma* were placed on CITES Appendix II. In some instances, such as the Belizean populations of the Mexican redrump, *Brachypelma* tarantulas are unquestionably common and widespread. In contrast, very little is known of the population status of some forms, such as the Mexican flameknee tarantula, *B. auratum,* which are restricted to small geographic areas and particularly vulnerable to overcollection. One negative result of the CITES listing has been decreased supply of some species, leading to increased desirability and thus demand, which has encouraged the development of a black market in smuggled specimens. While some tarantulas may be abundant enough to be impervious to unregulated collecting, many species have small distributions and are only abundant locally, and there is good reason to worry that the long term survival some of the more attractive members of this group could be threatened due to the extermination of local aggregations.

During the 1960s and '70s the Mexican redrump tarantula was a staple on animal dealers' price lists. Most of these specimens were distributed by a few of the largest companies based in the Miami area, and had been collected in Belize. Fortunately at the present time none of the tarantulas of Belize are being collected commercially, at least not on a large scale. Mexican redrumps are still frequently seen for sale in the pet trade but the majority of these specimens are captive-bred. Cinnamon tarantulas are also sold occasionally, but are not in high demand, perhaps because of their subdued coloration.

The tendency of tarantulas to cluster into small areas and being common only sporadically across the entirety of their geographic distribution leaves them vulnerable to natural stochastic events such as drought. Tarantula biologist William J. Baerg described a drastic decline of a large aggregation of the Mexican *Aphonopelma crinitum* over a period of 28 years. When he discovered the spiders living in an area of mesquite scrub on a ranch near Durango, they were the largest aggregation of tarantulas that he had ever seen. The size and density of this population must have been truly amazing. Across an expanse of eight ha (20 acres) the burrows were crowded as close as 61–91 cm (2–3 ft) apart. They may have numbered in the thousands. Baerg did not return to the site for 8 years, during which time the area suffered a severe drought. Upon his return he found only a few tarantulas remaining. Twenty years later they had all but disappeared. Baerg could only attribute the decimation of the species to the chronic drought. Most animals are able to migrate away from severe local conditions. Even plants can survive local extinctions due to their ability to disperse seeds to distant areas. As a group, spiders are one of the most proficient colonizers in the animal kingdom, but tarantulas are the glaring exception by completely lacking any ability to travel farther than their legs will carry them. While it is true that some males have been observed to move over distances of a kilometer or more, there is no evidence that females or subadults of either sex ever do. Tarantulas in Belize occur in discrete and sometimes isolated patches of abundance across the entirety of their distribution. When local environmental conditions become untenable, tarantulas are among the least equipped of organisms to move to a better location and would probably face extinction.

It appears that the most common form of tropical land altering activity by man poses little threat for most Belizean tarantulas. Despite the destructive connotation of the term "slash and burn," the small patches of land cleared for subsistence farming (known as milpas in Latin America) actually pose little or no threat to most tarantula species and may even offer more optimal habitat than the surrounding bush. The creation of a traditional milpa begins by felling all the trees and large shrubs in a 1–2 ha (2.5–5 acre) area of forest (Fig. 5.1). After the cut debris has dried, it is burned to enrich the soil with nutrients. The land is farmed for a year or two, then abandoned for a new location where the process is repeated. Once abandoned, the old milpa sites are not recultivated for decades, if ever, allowing full recovery of the former floral and faunal diversity. Most terrestrial tarantulas prefer open habitat with few trees and abundant sunlight reaching the ground. Especially during the first 4 or 5 years of fallow, milpas provide excellent habitat for tarantulas as indicated by the large populations typically found living in them.

Even the intense fires used to create milpas do not appear to harm tarantulas. Within a few days after the fires have died down and the ground has cooled, burrows become conspicuous due to the fresh lining of silk overlaying the ashes around the entrance, indicating the post-fire survival of the occupant. Juvenile tarantulas survive these fires as well. Large palm trees which have fallen prior to land clearing do not burn well due to their density and high moisture content. Many milpas contain several large, decaying palm trunks which have survived the fire and are too heavy for the farmer to remove. These logs are the favored refuge of a wide variety of small creatures, including juvenile tarantulas. Examination of these logs immediately after a fire reveals that the inhabitants have avoided any injury.

Increasing human population and the resulting pressure to use any available

Figure 5.1 Aerial view of a traditional milpa. (Photo by Scott Franklin)

land for crop production has changed the way subsistence farmers use the land. Traditional slash and burn practices are giving way to a more damaging type of activity known as migratory destructive agriculture, where fallow periods are greatly shortened. The destruction of the natural vegetation proceeds along an advancing front, while any land already put into cultivation is not permitted to undergo more than a few years of succession before being burned and planted again. This eventually results in a patchwork of cleared land and early successional scrub (Fig. 5.2). The forest never has an opportunity to recover. This type of destructive land use is becoming the norm throughout Belize and the rest of the Yucatán Peninsula. The large expanses of permanently cleared land that result soon become wastelands for most wildlife, including tarantulas.

There are, however, some species that do not appear to be adaptable to even small-scale man-made habitat alterations. The two Belizean species of arboreal tarantulas have not been found outside old growth forest. These spiders may require the high canopy, dim light, and high humidity treetop refuges that are rare or absent in other types of habitat. Another species of concern is Gutzke's tarantula, known from only one specimen discovered in an area where the tarantulas are better known than any other location in the Yucatán. This location, the Lamanai Archaeological Reserve, was set aside to protect the extensive Maya ruins and encompasses 385 ha (951 acres) of secondary forest approximately 80–100 years old (Fig. 5.3). The reserve is merely a fragment of what was once a continuous cover of forest across northern Belize. Today, the area surrounding the reserve is either cleared for agriculture or is wetland and thus unsuitable for tarantulas. Was Gutzke's tarantula a species adapted to the original forest

# Why Worry About Tarantulas?                                    27

Figure 5.2 Aerial view of migratory destructive clearing. (Photo by Scott Franklin)

habitat? It is likely that the rarity of this spider is due to its being a forest-adapted species in a region where none of the original forest exists. Gutzke's tarantula may be an unfortunate example of how the destruction of tropical

Figure 5.3 A vista in the Lamanai Archaeological Reserve. (Photo by Scott Franklin)

forests is occurring faster than biologists can explore these areas and describe the still unknown life forms they support. If the single Gutzke's tarantula had not been serendipitously discovered in 1995, the very existence of this spectacularly beautiful spider might never have been known.

# Chapter 6

# Tarantula Habitats

Regions can be categorized into biotic provinces in a variety of ways, according to the needs of the designator. Most often, vegetational zones are defined because floral character and composition usually reflect the subtle distinctions that climate and topography impart to a landscape, and which in turn help determine what wildlife is present or absent. Tarantulas, being sensitive to a different combination of environmental factors than many other creatures, require a slightly modified classification system.

Belize can be divided into seven vegetational habitat types that are relevant to tarantula species distributions: savanna, pine ridge, secondary forest, primary forest, strand forest, mountain pine ridge, and milpa. There are also two important elevation zones to consider: lowlands, which are less than 200 m (656 ft) and highlands, which are greater than 400 m.

## Savanna

Savannas are prevalent throughout the lower elevations of Belize. Large tracts of this habitat type are visible during a drive along the Western and Northern highways. This habitat is characterized by large, open expanses vegetated by grasses and sedges, with widely scattered pine (*Pinus caribaea*), oaks (*Quercus* spp.), palmetto palm (*Acoelorrhaphe wrightii*), and calabash (*Crescentia cujete*) (Fig. 6.1). The substrate in savannas is usually more sandy than in adjacent habitat. The ground is typically covered by tall grasses, making it difficult to locate tarantula burrows. Tarantulas often excavate their burrows at the base of bunches of grass. The sandy soil provides a good medium for burrowing and tarantulas can be very abundant. Some savannas are associated with wetlands and are flooded during the rainy season, yet still support many tarantulas. Savannas that are permanently wet and marshy are devoid of tarantulas. Studies are required to learn what happens to local populations of spiders after these areas become submerged. There is no known precedent among tarantulas to indicate that individuals are capable of vacating an area when the ground becomes saturated and then return months later when conditions improve, yet this appears to be what occurs. It is a puzzle how dense populations of adult tarantulas manage to persist in seasonally flooded savannas, such as the area around Dawson Creek on the eastern side of the New River Lagoon. A large species of fossorial wolf spider (*Hogna carolinensis*) also occurs in these areas and its burrows are similar to those made by tarantulas.

Figure 6.1 Savanna.

## Pine Ridge

This habitat resembles savanna, and is often adjacent to it, but is distinguished by a greater density and diversity of trees (in Belize, "ridge" is a term applied to a forested patch and does not imply higher elevation). Typical pine ridge habitat can be seen in the vicinity of August Pine Ridge Village in Orange Walk District, Belize (Fig. 6.2). The substrate here is quite sandy, sometimes composed exclu-

Figure 6.2 Pine ridge.

sively of sand near the surface. Pine trees dominate the sparse canopy, which creates patchy shade. A sparse to moderately dense midstory of smaller pines, oaks, and shrubs is also characteristic. An understory of grasses mixed with small, herbaceous shrubs, interspersed among scattered areas of exposed ground, makes finding tarantula burrows fairly simple.

## Secondary Forest

Most of Belize's forests have been logged at some time in the past. Even high canopy forests with very large ceiba (*Ceiba pentandra*), guanacaste (*Enterolobium cyclocarpum*), and cedar (*Cedrela odorata*) in some of the country's protected areas are actually secondary forest no older than 100 years. Various categorizations have been proposed for the hardwood forests of Belize, but these designations are largely unimportant for an understanding of tarantula distribution. Several species of theraphosid spiders occur in hardwood forest, but areas dominated by stands of cohune palm (*Orbignya* sp.), a common habitat type, are avoided by spiders due to the frequently saturated, muddy substrate. Tall trees visually dominate older secondary forest habitat and shade the forest floor (Fig. 6.3). As a result of the low light levels, tarantula activity may continue throughout the day and they may often be seen at the entrance of their burrows, waiting for prey, at any hour.

Tarantulas are not easy to find in heavily forested habitats. Low light levels and the abundance of ground cover and fallen objects make the few that do live there hard to see. The majority of Belizean species avoid shaded habitats and favor

Figure 6.3 Secondary forest.

more open locations which receive some sun. Exceptions to this rule are the Costa Rican orangemouth tarantula, *Psalmopoeus reduncus,* the Livingston tarantula, and the population of pygmy tarantulas living in the Cayo District, which all prefer forest habitat. One terrestrial microhabitat afforded by the forest that is exploited by tarantulas are the bases of large trees. Forest-dwelling tarantulas often dig their burrows at the bases of trees, especially along buttress roots.

## Primary Forest

Although most of Belize's original forest has been cut or severely fragmented, a few extensive undisturbed tracts still exist. The largest unbroken tracts of primary tropical forest are in the Toledo District (Fig. 6.4). A good indicator of whether a forest has been recently cleared is the abundance of very large mahogany trees (*Swietenia macrophylla*). Nowhere in the Yucatán were these trees passed over by loggers during the colonial period, so only in virgin or very old secondary forests will such trees be common. Arboreal tarantulas are most likely to be encountered in primary forest.

## Strand Forest

Many of the approximately 450 offshore cays that line Belize's barrier reef contain small strips of thickly vegetated areas that provide suitable habitat for tarantulas (Fig. 6.5). Dominant vegetation includes coconut palm (*Cocos nucifera*), sea

Figure 6.4 Primary forest.

Figure 6.5 Strand forest.

grape (*Coccoloba uvifera*), mango (*Mangifera indica*), and buttonwood (*Conocarpus erecta*). Mangroves (*Rhizophora* and *Avicennia*) are abundant over extensive areas along the coast and on cays, and indicate nearby soil that is too wet for tarantulas. Only the Mexican redrump tarantula is known to occur offshore, but even very small cays can support surprisingly large populations.

Tarantulas do not typically live in burrows in the cays. Perhaps due to the high water table, even large adults which would very rarely be found living outside a deep burrow on the mainland do not construct burrows in strand forest. Instead, all size classes tend to hide under the abundant plant debris and washed-up trash and boards that are so prevalent in this habitat. Here again, as with seasonally flooded savannas, the tenacity of the tarantula can be appreciated when one considers that the small cays are sometimes overwashed by the sea during tropical storms, yet the spiders remain.

The larger islands and coastal regions such as Ambergris Cay offer more diverse terrestrial habitat than tiny cays. Some of the interior sections of these land masses support forests which closely resemble those on the adjacent mainland.

## Mountain Pine Ridge

The highland region of Belize designated as Mountain Pine Ridge is distinguished by the oldest exposed surface rock known in Central America, and is the only portion of the Yucatán Peninsula never to have been submerged beneath the sea. Such a history promises the occurrence of relict species, particularly with poor dispersers such as the tarantulas. Yet, in spite of the antiquity and unique geological history of the Mountain Pine Ridge, no endemic theraphosid spiders have been found and only one taxon, the Cayo tarantula, *Citharacanthus meermani,* is known to occur there at all. Incongruous-looking stands of mountain pine (*P. oocarpa*) dominate the scenery (Fig. 6.6). The burrows of Cayo tarantulas are conspicuous because the white silk that radiates out from the opening stands out starkly against the reddish brown pine needles that blanket the ground.

## Milpas

Milpas, small plots cleared and planted in subsistence crops, are ubiquitous in Belize. Although not a natural formation, milpas are a distinctive habitat for tarantulas. A traditional milpa is a 1–2 ha (2.5–5 acre) clearing in the midst of brush or forest, planted with citrus, beans, corn, banana, plantain, or a mixture of these (Fig. 6.7). Between rows of crops the ground cover is sparse or nonexistent, and the surface is exposed to direct sun for most of the day. When left fallow, milpas quickly become overgrown with early successional invaders. As would be expected, leguminous species dominate. Plants such as cotton tree (*Cochlospermum vitifolium*), dog's tongue (*Physchotria tenuifolia*), polly red head (*Hamelia patens*), and various piper (*Piper* spp.) are typical. Tarantulas congregate in small milpas, especially in those adjacent to forest, making this habitat the best place to find tarantulas quickly and in large numbers. In stark contrast are the vast areas of cleared land due to migratory destructive agriculture, or where livestock is allowed to graze. Few if any tarantulas can be found in such areas.

Figure 6.6 Mountain pine ridge.

Figure 6.7 Milpa.

## Elevation

The most significant physical feature which correlates to tarantula distributions in Belize is elevation. Only one species, the Mexican redrump, seems oblivious to elevation and occurs countrywide. With the exception of the ubiquitous *Brachypelma*, areas below 200 m (656 ft) and areas 400 m (1311 ft) and higher contain theraphosid fauna that bear no resemblance to each other. A transition zone between 200–400 m exists in the foothills around the perimeter of the Belizean highlands where some lowland and highland species are sympatric. The Rio Bravo Escarpment in extreme northwestern Belize is not considered highland but is a significant enough feature to represent a dispersal barrier for at least one species.

# Chapter 7

# The Hidden Life of Belizean Tarantulas

One of the satisfactions of studying tarantulas is how easy it is to make significant contributions to our understanding of them. Most spiders are very conspicuous because they spend their time suspended in webs and are active during the day, making their lives easily accessible to anyone interested in them. As a result, the basic life histories of many species of web-dwelling spiders have been recorded and the focus of study has moved into areas that are beyond the technical expertise of most lay persons. In contrast, the life history details of even the most familiar species of Belizean tarantula, the common redrump, is filled with gaps. Through careful observation an amateur biologist can make important contributions to science and our knowledge of tarantulas. The abundant gaps in our understanding of tarantula natural history revealed throughout the following overview emphasize how many fruitful areas of study are available for any student of tarantulas.

Why do we know so little about tarantulas? One of the primary reasons is that these spiders lead such hidden lives. Almost every routine of life is conducted deep inside the burrow or inside silken tubes high in the trees. To make matters more difficult, tarantulas close their burrows at certain times, in effect "drawing the shades" during some of their most important and interesting behaviors. The world of the tarantula is not easily accessible to us unless we are willing to poke our fingers down these holes and use our inquisitiveness to dig out the facts.

## The Burrow

In terms of what sort of retreat they live in, the tarantulas of Belize fall into three categories. Arachnologist Sam Marshall coined the terms obligate and facultative burrower to distinguish between two modes of living by terrestrial species. Obligate burrowers are those which excavate burrows themselves and are always found living in them (Figs. 7.1, 7.2). The only time a tarantula that is an obligate burrower would be out on the surface is when looking for mates (if a mature male), or if the burrow was disturbed or destroyed, such as after very heavy rainfall or if dug up by a predator. Facultative burrowers sometimes excavate burrows themselves, but depending on the local environment they also utilize the abandoned burrows of other animals or simply secrete themselves under surface objects such as logs, rocks, leaf litter, or trash (Fig. 7.3).

Among Belizean species, only *Brachypelma* is a facultative burrower when adult. However, juvenile tarantulas of all terrestrial species are likely to be found living amongst surface debris or inside rotting logs (Fig. 7.4). Juvenile tarantulas will take up residence in whatever refugia they can find until they are able to excavate their own burrow.

Figure 7.1 Pygmy tarantula burrow opening.

Figure 7.2 Mexican redrump tarantula burrow opening.

Figure 7.3 Mexican redrump tarantula under debris.

# The Hidden Life of Belizean Tarantulas

Figure 7.4 A rotting log harboring juvenile tarantulas.

The third type of tarantula retreat is the kind utilized by arboreal species. These spiders have brought the burrow up into the canopy with them by spinning silken tubes that are similar in structure to the holes in which their ground-dwelling relatives live. A typical tube web consists of a cylindrical tunnel of silk with an opening at one or both ends, and is attached to the surrounding vegetation by a tangle of silk that may be very subtle or quite extensive. Some arboreal tarantulas are adept at camouflaging their tube webs by incorporating bits of vegetation onto the silk and disguising it. More typical of the two Belizean tree-dwelling species is to construct the web behind some object like a piece of loose tree bark or among the leaves of a bromiliad. In either case, the retreats of arboreal tarantulas are usually very hard to find. Another place that arboreal tarantulas often spin their tube webs is in the thatch of houses and especially, under the gaps of corrugated tin roofs (Fig. 7.5).

Tarantula burrows are not simply holes in the ground. Typically, they are lined with silk, especially near the opening, and terminate with an enlarged chamber where molting, egg laying, and egg incubation can take place in security. The pygmy tarantulas characteristically add a side tunnel off the main burrow, but the burrows of most Belizean tarantulas are simple tunnels. In Belize, burrows made by the larger species penetrate 30–60 cm (11.8–23.6 in) below the surface, while those of pygmy tarantulas are only 13–25 cm (5.1–9.8 in) deep.

Tarantulas spin a fine gossamer sheet of silk across the burrow entrance each morning prior to the cessation of their daily activity (Fig. 7.6). This is could be to reduce the stirring of air within the burrow and the entry of drier outside air, thus ensuring stable humidity and temperature. The veil of silk also prevents debris such as leaves and dirt from being blown down into the burrow. Yet another possible

Figure 7.5 The tube web of a Costa Rican orangemouth tarantula.

Figure 7.6 Silk sheet over burrow entrance.

advantage gained by this daily habit is to reduce the likelihood of predation by wasps, perhaps the primary danger to an adult tarantula, either by making it more difficult for the wasp to invade the burrow, or by reducing the release of chemical cues or scents which would alert the wasp to the presence of a tarantula.

It is normal to find a number of burrows in close proximity to each other (Fig. 7.7). These aggregations may be composed of a single species, but they are often made up of two or more tarantula species. Sometimes holes dug by wolf spiders are scattered among the tarantulas' burrows as well.

Figure 7.7 A burrow aggregation. Arrows mark individual burrows.

## Microenvironment in the Burrow

Tarantulas are extremely vulnerable to desiccation and heat stress. Despite the fact that many of their habitats in Belize are hot and xeric during the dry season, tarantulas live in a remarkably stable and mild environment by staying inside their burrows. In northern Belize, the average daily temperature inside tarantula burrows at a depth of 48 cm (18.9 in.) is approximately 25°C (77°F), and daily fluctuations generally stay within a 2° range (Fig. 7.8). Considering that the ground surface temperature adjacent to the burrow opening can range from 14°–40°C (57.2°–104°F), it is easy to see that tarantulas are virtually imprisoned inside their burrows for most of the day because exposure to the outside environment would quickly kill them.

Many burrows are excavated under the large limestone boulders that are typical of lowland habitats. Such burrows tend to be warmer than burrows located out in the open and away from large rocks. During the hottest hours of the day in late afternoon, burrows located under limestone boulders tend to be slightly warmer than burrows located out in open ground. At night, a tarantula burrow remains 2°–4°C (3.6°–7.2°F) warmer than the surface temperature, a difference significant enough to be noticeable when a finger is placed inside a burrow.

## Courtship and Breeding

The onset of tarantula breeding season coincides with a pulse of emergence by the mature males. The males start making their annual appearance in early Au-

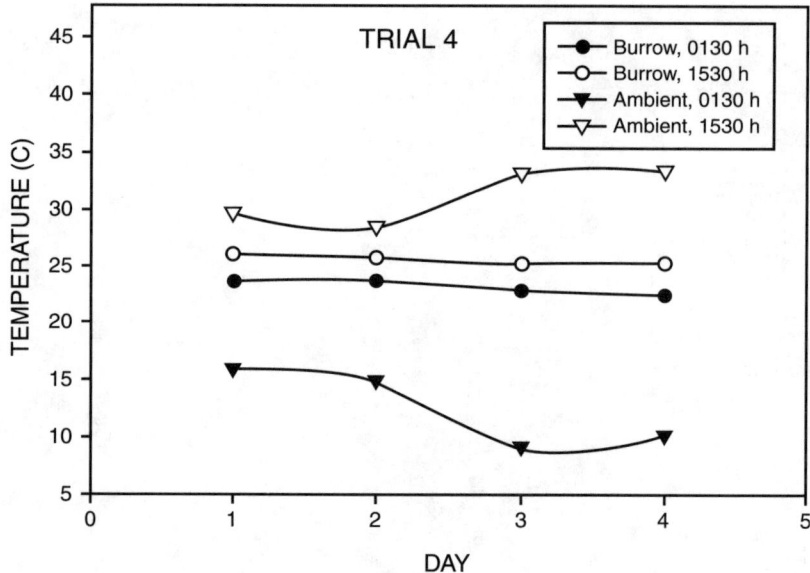

Figure 7.8 Graph showing burrow temperature vs. ambient temperature fluctuations.

gust, with peak abundance occurring sometime during September or October depending on locality, and are scarce by late November. With so many roaming males out at all hours of the day it can seem like Belize's forests, fields, and villages are full of tarantulas, and in fact they are, but this is the only time of the year that tarantulas go about their lives in a manner that makes them conspicuous.

It is interesting to speculate why the males emerge during such a narrow time period. The activity takes place during the end of the wet season, and the overcast sky and damp ground are just the right conditions to permit these easily desiccated and light sensitive creatures to take their excursions during daylight hours. It's hard to imagine a tarantula being able to expose itself for long during a typical afternoon in the dry season, and even at night the crackling dead leaves and hard dry ground would be an inhospitable environment for a tarantula.

It might seem that with so many males maturing within such a short time frame that intense competition for mates would be created. Yet constraining the mating period into just 3 months may actually offer some advantage. Once a female has mated she will be unable or unwilling to do so again for up to 9 months in order to oviposit, tend the cocoon, and guard her offspring until they disperse. If males matured continually throughout the year they would find many females already occupied with some aspect of reproduction or maternal care, and much effort and consequent risk of predation would be associated searching for a receptive mate. With all adult females having completed rearing the previous season's broods and ready to mate at the same time, all mature males start on a level "playing field," all other factors being equal. However, the other factors are never equal, which is the variation that natural selection plays off of. Under the natural scenario, the advantage lies with the males that possess characteristics that impart fitness to the population: large size, agile locomotion, better chemosensory perception, and superior defensive capabilities. Up to a point, a male simply emerging earlier than other males would find himself at an advantage by having the entire population of receptive females virtually to himself. However, the system offers no advantage to extremely early emergers because they would be roaming a landscape filled with females still occupied with cocoons and unreceptive to breeding. Males which emerge late in the season have fewer potential mates still awaiting a suitor. Thus the sudden and brief duration of mature male emergence could be the result of stabilizing selection due to negative fitness consequences for males at either extreme of emergence timing, and limits fitness advantage to males possessing qualities that benefit both sexes and the species as a whole.

Only one study has exploited the advancements in radio transmitter miniaturization in order to follow the movements of mature male tarantulas. Margaret E. Janowski-Bell found that males of a Texas species travel 1 kilometer or more in search of mates. It is not known whether any Belizean tarantulas move as far, but unless inbreeding is occurring within aggregations it is likely that they do, because clusters of burrows are often widely separated from each other. One difficulty with radiotelemetry studies of tarantulas is the problem of maintaining adhesion of the transmitters to the spiders. The only practical attachment point is the carapace, but even the most durable glues detach within a few weeks at best, rendering long term monitoring problematic. An alternative is being developed, using tiny passive-integrated transponders which are surgically implanted into the abdomen, providing a permanent marker. At one location in Orange Walk District over 100 Mexican redrumps are carrying these microchips and are part of a multiyear life history study.

The males spin a sperm web about a week after their ultimate molt. This activity is performed in the burrow, prior to beginning the search for females that will be the male's overriding purpose for the brief remainder of its life. If the male survives long enough to mate with a succession of females, additional sperm webs may be made, always under the protective cover of some object. Sperm web construction is one of the secretive behaviors of tarantulas that keep their lives hidden from our eyes. The process has rarely been described for tarantulas living in the wild, and never in Belize, but has frequently been observed in captives with no place to hide. The sperm web of the Mexican redrump tarantula is rectangular in shape and approximately 2 cm × 4 cm (0.8 in. × 1.6 in) (Fig. 7.9). The web is suspended over the ground, attached to the side of the container at one end approximately 2 cm (0.8 in) above the ground so that it rises obliquely at one end. This creates a space beneath the web that the male can slip under, which he does on his back. After a moment of rest while lying on his back between the web and the ground, the male deposits a drop of seminal fluid onto the underside of the web. The fluid comes from the genital opening located within the epigastric furrow that is clearly visible on the anterior underside of the abdomen. It is beyond the epigastric furrow that the entry to the female's reproductive organs lays and into which the male deposits his semen. After producing the seminal fluid, the

Figure 7.9 A sperm web being constructed.

male crawls out from under the web and positions himself above it. He begins dabbing the tips of his palpal bulbs (see Chapter 10) into the semen, drawing it up into the bulb by capillary action. The male is now equipped to begin searching for a receptive mate. As soon as the male has finished using the sperm web he destroys it.

Published descriptions of tarantula courtship and mating in the wild are scarce, but these events are sometimes observed by biologists working in the field. On one occasion while in Orange Walk District, I watched a male cinnamon tarantula try to coax a female out of its burrow in order to mate. The male had already found the entrance to a female's burrow when I first noticed it. He performed a series of rapid taps with his first pair of legs while simultaneously moving his pedipalps up and down. This behavior continued for about 3–4 seconds, whereupon he ceased moving his appendages and began to shiver. Most of the movement responsible for this trembling behavior appeared to be generated at the junction of the legs and body, but the vibration involved the entire spider. After a few seconds, the male moved a step or two closer to the burrow and repeated the cycle of leg and pedipalp movements alternated with shivering. I watched this process continue for about 5 minutes and then captured the male. This specimen became the holotype for the taxon. On later occasions I placed mature male cinnamon tarantulas at the entrance to females' burrows and watched the female tap back and ascend to the surface. Although no sounds were audible to my ears, the movements of the male, particularly those of the pedipalps, resembled the motions associated with stridulation by other theraphosids. It might be fruitful to study these movements with sensitive recording equipment to determine if tarantula courtship involves the production of sound.

For most species, mating is preceded by a similar sort of "drumming" which may be done with the two front legs, the pedipalps, or both (Fig. 7.10). Drumming communicates the male's presence to the female, who if receptive will come to the burrow entrance and respond with a drumming signal of her own. The drumming pattern of each species appears to be different and probably serves as an isolating mechanism between species, in much the same way as the different breeding calls of frogs and toads do.

After emergence from the burrow, a receptive female will lean back on its rear legs and raise its foreparts to expose its genital region (Fig. 7.11). With the exception of Gutzke's tarantula, mature males of all Belizean species possess spurs on the underside of the tibial segment of the front legs, which they use to hook under the female's fangs and push her back further so that intromission can be achieved (Fig. 7.12). Intromission consists of the male inserting a sperm-loaded embolus into the genital opening of the female, where the sperm is released. Sometimes only one embolus is used but often both emboli are inserted in succession. After successful insemination the male disengages its tibial spurs, allowing the female to resume her normal position (and her ability to attack the male) and quickly retreats. In captivity, the males of some species rarely escape being consumed by the female after copulation, while in other species cannibalism is uncommon. Whether males are often eaten by females in nature is not known. If the male survives the encounter, it may spin another sperm web and recharge its bulbs with more semen. A male mates with as many females as possible dur-

Figure 7.10 A male Mexican redrump tarantula drumming at a female's burrow.

Figure 7.11 Courting Mexican redrump tarantulas.

ing its short time as a sexually mature animal. In the wild, most males probably die when their conspicuous wanderings bring them to the attention of predators. In captivity, the males will spin several sperm webs over a period of several months and mate with a succession of females if given the opportunity. These captives

7.12 Mexican redrump tarantulas mating.

gradually begin to decline in vigor, refuse food and become emaciated, and usually die in about 4 months.

The seminal fluid that the male leaves behind can be stored in the spermatheca for an extended period of time. In some species this structure is bilobed and takes the form of two separate receptacles (Fig. 7.13), while in others it is a single compartment. Here the sperm remain until the eggs are laid. As each egg moves past the lumen of the spermatheca on its way out of the female, it comes in contact with spermatozoa and is fertilized. This arrangement gives the female some flexibility in timing by allowing a fortuitous pairing with a fertile male and copulation to occur even if her physiological state or the environmental conditions are not suitable for egg laying. She can mate when the opportunity presents itself and be somewhat free to lay her eggs at a later time when the survival of her offspring, and herself, is more likely. It is not known how long the sperm will remain viable in the spermatheca after mating, but it is certain that the female must lay eggs before the next molt or the breeding will be ineffectual. This is because the lining of the spermatheca, along with any spermatozoa they may still contain, are cast off along with the rest of the exoskeleton.

## Oviposition

The cocoon (also called the eggsac or ootheca) that female spiders fashion to contain and protect their eggs is an important function of tarantula silk. A series of stereotyped behaviors associated with cocoon construction has been observed

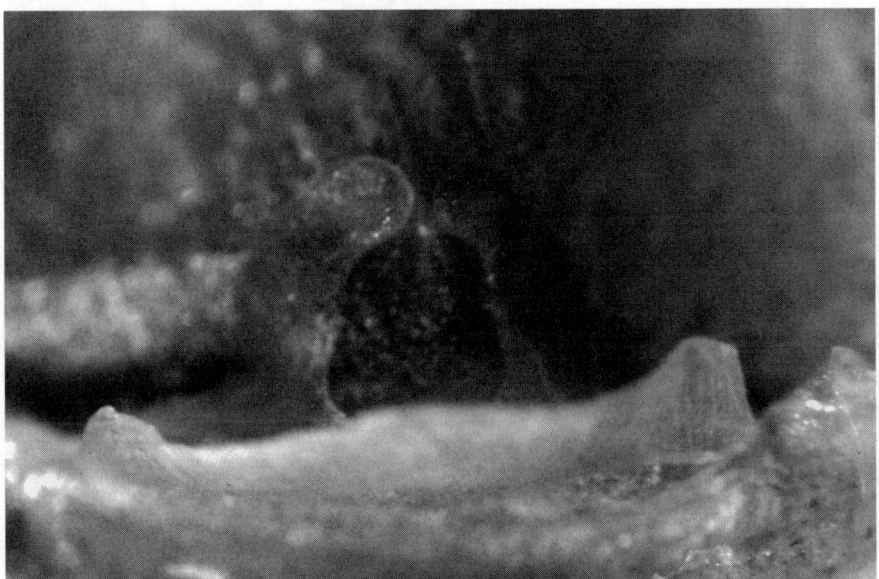

Figure 7.13 Spermathecae, bilobed.

in Mexican redrump tarantulas. The female spins a heavy mat, roughly circular in shape and slightly depressed in the center, creating a shallow bowl. The silk is thick and spongy and feels different from other silken structures spun by tarantulas. The eggs, resembling yellow caviar, are laid to accumulate in a pile at the center of the mat. The female spins a cover over the egg mass and fashions the flat and flabby sack into a sphere by a series of turns, pokes, and tucks until the cocoon is about the size and shape of a golf ball. Oviposition in the arboreal species has not been as closely studied.

## Maternal Care

In terrestrial species, females carry the cocoon with their fangs and pedipalps, setting it down only briefly to turn it, or occasionally, to take prey. Some females make an additional investment in the well-being of their offspring by incorporating irritating body hairs (described later in this chapter under "Predators and Defenses") into the fabric of the cocoon (Fig. 7.14), which discourages egg predation by parasitic scuttle flies (family Phoridae).

After the eggs are laid, some females will block the entrance of their burrow with a soil plug (Figs. 7.15, 7.16). This is accomplished by scraping the sides of the burrow wall with the fangs and binding the fragments together with silk to form pellets of dirt, which are then carried to the surface and stuffed into the entryway. In the United States, tarantulas have been observed doing this just prior to the onset of winter, and these species do not reopen the burrow until spring.

Figure 7.14 Comparison of Mexican redrump and cinnamon tarantula cocoons.

# The Hidden Life of Belizean Tarantulas

Figure 7.15 Pygmy tarantula burrow cutaway. Note soil plug near surface and juveniles in burrow.

Figure 7.16 Occluded burrow entrance.

Tarantulas studied by Mandy Kotzman in Australia plugged their burrows to prevent dehydration during dry weather. For Belizean species, burrow occlusion seems to be primarily associated with ecdysis. Although some females plug their burrows prior to egg laying, others will carry the cocoon up to the surface for reasons which are not known.

Female tarantulas provide a considerable amount of care to both the cocoon and the spiderlings after they have emerged. Rarely throughout the 8 weeks required from the time of oviposition to emergence of spiderlings from the cocoon will the female set her eggsac down (Fig. 7.17). If she does it is only briefly, either to turn the bundle and let the eggs or spiders on the bottom be at the top for awhile, or so she can take an occasional prey item. Female tarantulas feed only sporadically, and sometimes not at all, while tending an eggsac.

Passive feeding by juvenile tarantulas on prey captured by their mothers has been observed in some tarantula genera. On the Caribbean island of Sint Eustatius, burrows of *Cyrtopholis media* often contain several large juveniles in addition to an adult female. These spiders are probably behaving similarly to some African tarantulas (*Hysterocrates* spp.) in which the female captures large prey and tolerates offspring as they crawl over her body, gaining access to the masticated item and feeding from it. Belizean tarantulas do not appear to engage in this form of maternal care. In contrast to the species just mentioned, the burrows of Belizean tarantulas have never been observed to contain more than one individual. Spiderlings disperse en masse within a few days of emergence and are never found sharing adults' burrows at later times.

Figure 7.17 A Mexican redrump tarantula tending her cocoon.

## Life Inside the Cocoon

Little is known about the life of the tarantula while still in the cocoon. Although termed an eggsac, the eggs actually hatch in about 2 weeks, and for most of the time the female is tending a sac full of spiderlings, not eggs. With the exception of the pygmy tarantula, under natural conditions the spiderlings emerge about 8 weeks after the eggs are laid. Within the confines of the cocoon, hundreds of tiny spiderlings live piled on top of each other in a dense mass. The female keeps the spiderlings continually churning by frequently turning the sac, so that the spiderlings at the top of the pile are always changing.

While still in the cocoon, the spiderlings exhibit a tenacious bond with the silk. If a cocoon is opened prior to emergence, the spiderlings can be spilled out onto a dish and observed. If a small piece of the cocoon is cut off and laid beside the exposed spiderlings, they will quickly climb back onto it. A similarly textured material, such as tissue paper, will not elicit the same response. Perhaps the juveniles recognize the structure of the silk through tactile cues, or they may be responding to some scent imparted onto the cocoon by themselves or the female. During this period, the juveniles may also form a bond of sorts with their siblings. In the close quarters, they may develop a recognition of each other through chemosensation. Although studies testing the possibility of kin recognition in spiders have found no evidence of it, the remarkable behavior exhibited by some species of theraphosids immediately after emerging from the cocoon suggests that for an unknown period of time, tarantulas are able to recognize their kin and have a strong affinity for them.

Many mysteries surrounding the life of juvenile tarantulas while still inside the cocoon remain to be explored. For example, oophagy by postembryos has been observed in some species of typical spiders. Trophic eggs (infertile ova that provide a nutritious food for developing offspring) are a strategy that some female animals employ to better insure the survival and prosperity of their offspring. Perhaps female tarantulas provide this resource to their offspring, one of the many secrets that occur inside the cocoon. The behavior of the spiderlings within the cocoon is one of the most interesting and unexplored topics available to the student of tarantulas.

## Juvenile Dispersal

About 48 hours before the juvenile tarantulas leave the female's burrow, they migrate toward the surface and congregate just inside the entrance. For the first time in their life they begin using silk extensively, each contributing to a loose mat spun around the congregation point. There they wait until nightfall to venture out into the new environment and begin their solitary lives. However, before separating from each other they begin their new life in an interesting way.

The spiderlings of the Mexican redrump tarantula disperse in groups. Whether this behavior occurs in other species is not known. Sometimes during a single evening, but more often over the course of two nights, the juveniles begin dispersing from their maternal burrow in groups of 20–100, and walk through the forest in single file columns (Fig. 7.18). They have been observed to traverse up to 10 meters in this way, a great distance for tiny, slow moving creatures. They may travel much farther, but observations for long periods are difficult when the spiderlings move into areas where the ground is blanketed by deep leaf litter. This event takes place in every burrow containing spiderlings over the course a short 7–10 day period, with some variation in timing at a countrywide scale, probably in response to some local environmental cue. It is a fascinating experience to happen across one of these processions as it snakes its way across a forest trail. From a distance and caught in the beam of a flashlight, the line of spiderlings resembles a column of ants, or perhaps a slender snake. If approached too clumsily or if a bright flashlight is shone directly on the column, the spiders will scatter, but they quickly regroup into single file and proceed as before once the disturbance ceases.

It is possible that this strange behavior explains why tarantula burrows are so often found in dense aggregations. Perhaps the spiderlings walk until they find a microhabitat to their liking—someplace damp and with places to hide—and then all settle down into one small area. Alternatively, the procession may wander until morning approaches and the rising temperature and light level force them to establish residence wherever they find themselves at that moment. In either case, the implication is that the aggregations of adult tarantulas so typical in Belize are composed mainly or entirely of siblings. If correct, this hypothesis would also explain the need for mature males to wander during the mating season as a mechanism to avoid inbreeding.

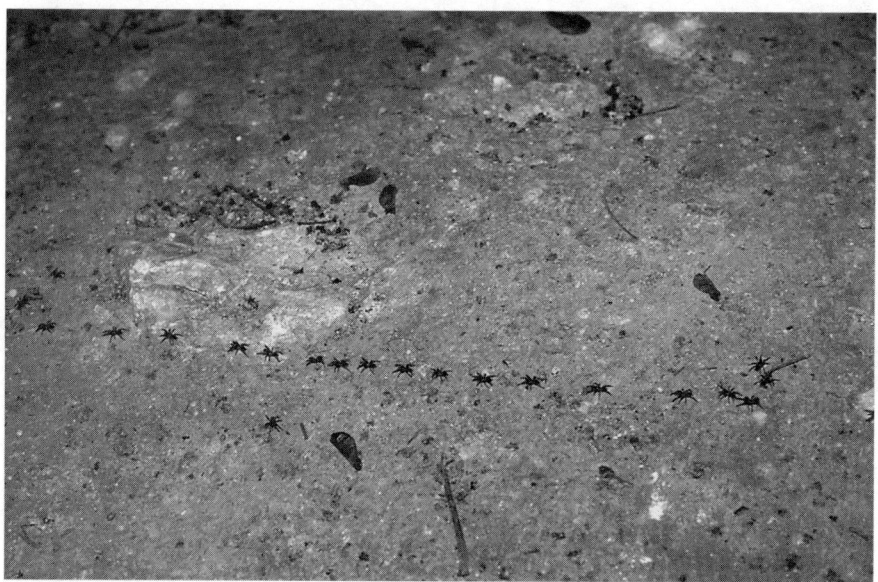

Figure 7.18 Aggregative dispersal of juvenile Mexican redrump tarantulas.

## Feeding Behavior and Diet

Tarantulas can be observed passively foraging for prey at night. Most individuals do not leave the burrow, but wait motionless at the entrance with the tarsi of their pedipalps and first two pairs of legs resting lightly around the rim or just outside the burrow (Fig. 7.19). Sensing vibrations created by the approach of a potential prey animal, the tarantula will wait until the creature is within 2–5 cm (0.8–2 in) or until it makes contact with one of the spider's legs before it seizes it. Although some fossorial spiders apply lines of silk that radiate out from the burrow entrance to serve as trip lines, these do not appear to be present around the burrows of Belizean tarantulas.

It is difficult to ascertain specifically what type of prey tarantulas prefer, or if there is any niche partitioning among sympatric species. Tarantulas are sit-and-wait predators and most species are probably generalists with regard to diet. Exceptions may be found in extremely large or small species, due to the availability of fewer prey of suitable size or the ability of the spider to exploit an untapped resource by virtue of its unusual dimensions in comparison to competitors. Although the pygmy tarantulas are small for mygalomorphs, they are well within the normal size range of most spiders and are probably unremarkable in their dietary preference, and no unusually large species occur in Belize.

In captivity, large Mexican redrumps will readily kill and consume preweaned mice and will occasionally take small adult rodents. While these species are certainly capable of consuming small vertebrates in situ, they likely have few op-

Figure 7.19 Tarantula waiting for prey. (Paint on leg is for individual identification.)

portunities to do so. Most native lizards are diurnal and thus not active when tarantulas are passively foraging. However, the terrestrial and nocturnal Yucatán banded gecko (*Coleonyx elegans*) and several small species of snakes are potential prey for large Mexican redrumps and perhaps cinnamon tarantulas as well (Fig. 7.20). I once observed a tarantula in the United States, smaller than an average size adult Mexican redrump, eating a large lizard that it had captured, so it is certainly possible that the larger Belizean species consume vertebrate prey on occasion.

By examining the discarded remains of prey in and near the burrow, it is apparent that large beetles are a common prey item of terrestrial species. This observation may lead to biased conclusions however, because arthropods lacking the heavy, chitinized elytra of beetles are more fully consumed and thus leave fewer identifiable remnants.

## Predators and Defenses

Few predators of tarantulas have been documented. One well-known tarantula specialist is the wasp known as the tarantula hawk (family Pompilidae). These solitary insects selectively prey on large spiders and paralyze them with their sting, whereupon they carry the spider to a burrow and entomb it, still alive, along with an egg deposited on the spider's body. After hatching, the larval wasp feeds on the immobilized spider. These wasps occur wherever tarantulas are found and are a frequent sight in Belize. There is a correlation between the size of the local spiders and the local wasps. An area with a large tarantula species will have a cor-

Figure 7.20 A large Mexican redrump tarantula feeding on a lizard.

respondingly large species of wasp. Small tarantulas predict small wasps. The largest tarantula hawk in Belize, which can subdue and carry off adult female *Brachypelma,* is a very impressive looking insect—deep indigo blue with brilliant scarlet wings, and up to 8 cm (3.1 in) long.

Other likely predators of tarantulas are small mammals such coatis, raccoons, cats, and skunks. Birds probably take an occasional spider found out in the open. The primary tarantula defense against mammalian attack is not its venomous bite as is commonly believed. Most tarantulas in the Americas are equipped with urticating hairs which are used to defend against vertebrate predators. Some species, including the cinnamon tarantula, also incorporate these hairs into the silk of the eggsac, where they create a protective barrier against certain predaceous fly larvae. In most species these hairs are located in a circular patch on the rear of the abdomen. In adults, the outline of this patch can be discerned because the overlying long setae are less abundant in this area. The patch is clearly visible on newly emerged spiderlings because they have very few setae of any other kind on their abdomens, resulting in a conspicuous black spot. The fact that the juveniles, although otherwise much less hirsute than adults, have a thick patch of these irritating setae indicates what an important defensive weapon they are for tarantulas of all ages. Each seta is armed with sharp barbs and upon contact with mammalian tissue becomes imbedded and elicits an allergic reaction. The allergic response varies according to the sensitivity of the victim, the tissue involved, and the specific type of irritating seta causing the reaction. Intense itching or burning, which can last several days, is the most common symptom, and this can be accompanied by urticaria. Six basic forms of urticaceous setae have been identified, each with different irritating capabilities and uses, termed types I–VI. Type I hairs are short (less than 0.6 mm), somewhat kinked and with a double row of backward-pointing barbs distally and a double row of forward-facing barbs at their base. Type II hairs (up to 1.5 mm) are found on some arboreal tarantulas, but not those living in Belize. Type III hairs (up to 1.2 mm) have many sharp, backward-pointing barbs and are the most irritating kind to mammalian tissue. Fortunately no Belizean tarantula possesses type III hairs. Also absent on local species are type IV hairs, which are extremely small (less than 0.2 mm), weakly barbed, and dartlike in appearance. Type V urticaceous hairs are unusual because they are found on the inner surface of the pedipalpal femur rather than the abdomen, and are known to occur on only the South American genus *Ephebopus*. Like type IV hairs, type VI setae are short ($\sim$ 0.3 mm), and are also very similar in morphology. Types I, III, IV, and VI are rubbed off the abdomen by the tarantula, using a rear leg in a repetitive kicking motion (Fig. 7.21). The hairs are so small and light they readily waft into the air and drift with the slightest breeze, some landing on the intended victim's skin or eyes, or inhaled to become lodged in the nasal passageways. Similarly, type V hairs are shed by the spider in defense, but by a different mechanism owing to their location. Type II hairs are not cast off into the air, but are incorporated into the fabric of the tube web by the arboreal *Avicularia* spp., producing skin irritation when the structure is touched.

Figure 7.21 A tarantula dislodging urticaceous setae.

In my experience, the hairs of Mexican redrump tarantulas produce the most severe rash, while those of cinnamon, Cayo, and Livingston tarantulas are capable of producing only mild irritations. However, human sensitivity to these irritating hairs varies among individuals. The usual result of skin exposure is a localized rash and hives that last from 1–2 days. Some people are more sensitive and develop persistent, burning skin irritation, and they should be very cautious about handling any tarantula without gloves. Do not inhale urticaceous hairs, or allow them to contact the eyes. These hairs can float directly into the eyes if your face is too close to a tarantula when it flicks off the hairs (a good reason not to let tarantulas crawl around on your face, which people sometimes do when showing off). The hairs can also become lodged in the eye after rubbing them with hands that have recently handled a tarantula. The irritation caused by these hairs will be severe in the eyes or on any mucous membrane such as in the nasal passageways, and could cause serious damage.

The bites of all tarantulas found in Belize and surrounding areas are of little consequence to most people. In fact, it's very hard to get a tarantula to bite at all, and when one is harassed to the point of biting, the symptoms are minimal. Bites produce a mild sting, but this feeling disappears after a few minutes. I was once bitten by a Mexican redrump when I inserted my finger into a burrow to determine which way it turned. My intrusion was answered with a bite on the tip of my index finger. A stinging sensation ensued, much like getting lemon juice into a shallow cut. Within a couple of minutes the stinging disappeared and I felt no other symptoms. On another occasion, I purposefully provoked a cinnamon tarantula to bite me on my forearm because I wanted to determine if the venom of this species was more toxic than the redrump's. This bite produced a soreness around the site of the bite that felt like a bruise and persisted for about 24 hours, in addition to the brief stinging sensation. Occasionally, the area around the bite will feel a little sore, and be tender to pressure, probably more as a result of the puncture wound created by the fangs than from any effect of the venom itself. A current tetanus vaccination and cleaning the bite with an antiseptic are the only treatment normally required for persons not hypersensitive to the venom. However, just as with the bites and stings of other venomous arthropods, some people may be allergic to the venom of tarantulas, or could develop an allergy later as a result of repeated exposure. In these cases, a normally harmless bite could have serious or even fatal consequences. Anyone with severe allergies, particularly those involving bee, wasp, or ant stings, should consult a physician about medications that should be carried before exposing themselves to the possibility of a tarantula bite.

## Longevity

Tarantulas live a long time, especially when compared to most spiders which seldom live longer than 2 years. Captives of some species have lived over 20 years. Most large Mexican redrumps seen in the wild are at least 10 years old. Yet another compelling reason never to harm a tarantula is the fact that any adult you find has already managed to avoid a multitude of hazards and survived for many years, a feat which surely deserves your kind treatment and respect.

Growth and maturation has been studied in only three species, none of them from Belize, but some generalities apply. Tarantula biologist William J. Baerg determined that a temperate-zone species, *Aphonopelma* sp., required approximately 10 years to reach maturity. The goliath birdeater spider, *Theraphosa blondi*, was studied by Sam Marshall and found to reach maturity (at least the males) in 3–4 years. These studies agree with numerous observations by tarantula hobbyists on captives in suggesting that temperate-zone species mature later and live longer than Belizean and other tropical species.

# Chapter 8

# How to Find Tarantulas

Tarantulas are everywhere in the Belize—below you, above you, maybe even right beside you. Visitors are frequently not aware of this fact. To learn about tarantulas in the wild requires the ability to find them. Armed with some knowledge of their habits, it's easy to start finding tarantulas, and with just a little practice coupled with the knowledge that these misunderstood spiders are actually shy and virtually harmless, it's not difficult to extract them from their retreats for closer examination. One advantage that tarantula hunters have over naturalists pursuing most other types of organisms is that these spiders don't move around much. Tarantulas remain in the same spot for weeks, months, and often years so that once located with a practiced eye, they can be revisited and observed repeatedly without additional effort.

The first encounter most people have with a tarantula is with the vagrant male Mexican redrumps. Persons visiting Belize during the first 6 months of the year are often completely unaware of the presence of tarantulas due to the absence of mature males. However, it's hard not to notice them during the breeding season, as the males are commonly found crossing roads or trails. Since some of the most distinctive species characteristics are mature male structures, these specimens are a good introduction to the nascent tarantula student because identification is so straightforward and unambiguous. While the wandering males are hard to avoid during the right time of the year, finding females and immature males requires more effort.

**Where to Look**

As a result of the tendency of tarantulas to occur in aggregations, every experienced tarantula hunter knows that when one tarantula burrow is found others are sure to be close by. This makes searching for tarantulas enjoyable and exciting. A search may go on all day without success, but the hunter knows that at any moment a big payoff may be at hand, and anticipation builds. Finally, after many hours, a burrow is found, and then within minutes, a dozen, several dozen, or in some cases a hundred or more burrows are quickly located. A search for burrow aggregations should be focused on sunny, gently sloping hillsides, particularly those with exposed rocks (Fig. 8.1). When hunting on flat ground, try to find a spot that has some incline, however slight. The most productive areas are often in small milpas surrounded by forest, alongside trails (Fig. 8.2), or along forest-field interfaces. Finding tarantulas in deep forest is much more difficult and there are, in fact, fewer tarantulas there. The best place to look in the forest is in small gaps created by tree falls (Fig. 8.3) or alongside trails. Since juveniles do not usually

Figure 8.1 Hillside habitat.

Figure 8.2 Trailside habitat.

excavate burrows until they have gained some size, look for these under rocks, boards and other surface debris, as well as inside rotting logs where they are sometimes very abundant (Fig. 8.4).

Caves, which are prevalent in the highlands, may harbor interesting tarantula species (Fig. 8.5). There are many, possibly hundreds, of caves in Belize. They

Figure 8.3 Forest gap habitat.

Figure 8.4 Digging through debris for juvenile tarantulas. (Photo copyright Sam Fried)

are distributed throughout the upland regions and are particularly numerous in the karstic limestone formations that surround the highlands. The majority of caves that have been biologically surveyed do not harbor tarantulas. In fact, many are virtually devoid of any permanent animal residents. Only caves which have suit-

Figure 8.5 Cave opening.

able humidity, temperature, and ventilation are "living" caves, and these may support whole ecosystems. Several of the most accessible caves are popular tourist stops and are frequently disturbed by human traffic. More remote and obscure caves offer a better likelihood of finding tarantulas. One such cave located in the Roaring Creek Valley is known as the Labyrinth of the Tarantulas. The small and unspectacular appearance of the mouth of this cave belies the abundant life within. Caves such as the Labyrinth of the Tarantulas often support populations of arboreal *Psalmopoeus* tarantulas. When exploring caves, search along the walls for evidence of silk retreats hidden within the limestone pockets (Fig. 8.6).

## When to Look

Time of day affects the ease with which tarantula burrows are seen. Early morning and late afternoon hours are the best times for spotting the burrows due to the shadows and the glint of reflection on the silken entrance veil that the lower angle of the sun produces. Time of year is also a factor. Although tarantulas are best caught after nightfall, locating burrows in the evening is difficult. In most locations, only a few daylight hours are needed to find enough tarantula burrows to keep a hunter busy catching spiders well into the night.

The burrows of pygmy tarantulas are quite hard to find due to their small size. The holes are no bigger than what would be made if a pencil were pushed into the ground, and a little vegetation or debris can completely obscure the entrance. Often it is necessary to simply choose a likely location and search the ground care-

# How to Find Tarantulas

Figure 8.6 Cave habitat.

fully on hands and knees until a burrow is found. Fortunately, as mentioned earlier, once one burrow is discovered a close inspection of the immediate vicinity will usually yield a good sample.

During the dry season, the herbaceous understory that develops in cleared areas, such as milpas, dies back. As a result, burrow openings are not hidden from view as they are during the rainy season when knee-high grass obscures the ground. This is also the time of year when farmers burn their milpas, and a few days after such fires the burrows stand out conspicuously to even the untrained eye. The best months to find tarantula burrows in Belize are January through May, after the onset of drier weather.

## What to Look For

Not every likely looking hole in the ground will be home to a tarantula. Numerous other small animals live in holes that resemble tarantula burrows. Tarantula burrows range in diameter from 1–7 cm (0.4–2.8 in), depending on the size of the spider. Holes larger than 10 cm (3.9 in) are most certainly rodent burrows. Wolf spiders live in burrows which look like those made by tarantulas. However, wolf spider burrows have somewhat ragged edges and the soil around the margin of the burrow entrance is loose, in contrast to the tarantula burrow which has a very smooth and circular entrance and is firmly packed around the edge of the opening. A second distinguishing characteristic of some wolf spider burrows is a collar of silk around the opening of the hole, which resembles a funnel (Fig. 8.7).

Figure 8.7 Silken collar around entrance to wolf spider burrow. (Photo by Gail E. Stratton)

This silk collar is only seen around burrows made by the larger species of wolf spider, but since these are the ones most likely to be confused with tarantula burrows, it is a helpful clue.

Another way to rule out wolf spiders is the presence of eye shine. Wolf spiders, having larger and more sensitive eyes than tarantulas, and also owing to the forward-directed placement of these eyes, reflect a pronounced greenish sparkle when a flashlight beam is directed at them. Just go out into the fields or forests on any night and slowly pan a flashlight back and forth across the most distant ground; the multitude of eyeshine revealed will attest to the amazing abundance of wolf spiders. Any large spider sitting at the entrance of a burrow and exhibiting this phenomenon is a wolf spider, not a tarantula. However, the two species of *Citharacanthus,* the Cayo and Livingston tarantulas, do seem to reflect a faint light from their eyes, an interesting quirk but in no way approaching the brilliant shine seen from a distance that wolf spiders produce.

Not every tarantula burrow is occupied. Perhaps because the former resident was carried off by a predatory wasp, or due to other factors still not understood, an occasional burrow will be found empty. There is a clue to look for which will reveal whether or not a burrow is occupied by a tarantula. All Belizean species line their burrows with silk. Sometimes this lining is hard to see, but if a thin twig is lightly dragged around the edge of the burrow opening the threads of silk will become obvious as they and some of the soil attached to them are snagged. Any burrow with silk incorporated into the soil around the entrance is sure to have a tarantula in residence. In addition, tarantulas are meticulous about keeping their burrows neat and clean, so the presence of a few dead leaves or grass, or some loose bits of soil down inside a burrow will usually indicate that the former resident is no longer present.

One important caution is necessary. Great care should be taken before fingers are poked down a potential tarantula burrow, because a number of creatures— some far more dangerous than any tarantula—often take up residence in abandoned tarantula burrows or excavate their own holes which closely resemble tarantula burrows. It's possible to encounter scorpions, wasps, and venomous snakes (particularly the mildly venomous black-striped snake, *Coniophanes imperialis*) in these burrows, so always visually confirm the kind of animal living down a hole before exposing yourself to a bite or sting (Fig. 8.8).

Once it's determined that a burrow is indeed occupied by a tarantula, the exciting question can be asked, "What kind of tarantula is living down there?" During the daytime tarantulas spend most of their time deep in their burrows and won't be visible. Sometimes during overcast weather, or if a burrow is situated in a shady location, the tarantula will be near enough to the surface to be seen and identified. In most cases, however, a better view can be had at night, when the tarantula is active.

Burrows found during the daytime should be marked so they can be easily located again after dark. Bright orange surveyor's tape works well for this purpose. After nightfall, if the burrow is approached very stealthily (because tarantulas can sense the vibrations of foot steps and will scurry to the bottom of their burrows at the slightest alarm), the resident spider will usually be visible resting partially out-

Figure 8.8 Scorpion in abandoned tarantula burrow.

side the entrance of the burrow with enough of itself exposed to allow identification. It cannot be overemphasized how softly the approach must be to get close to a burrow without alarming the occupant and sending it fleeing to the bottom. I am frequently amazed at how faint a disturbance tarantulas can detect. They seem to be interwoven into the fabric of the earth itself, their capacity to sense ground vibrations is so remarkable.

## Collecting Techniques

In some cases it will be necessary actually to have the spider in hand to make a close examination before positive identification is possible. If actually catching a tarantula is the goal, there are two techniques which can be used. It seems that everywhere in the tropics, young boys and girls learn how to "fish" for tarantulas. There's no better technique available, even for the professional tarantula biologist, and it's good sport as well. First, a burrow must be successfully approached without unduly alarming the resident spider. A stiff blade of grass or a thin twig is inserted down the burrow, then wiggled and twirled in a manner meant to mimic the vibrations that a large insect would create (Fig. 8.9). Tarantulas are easily fooled by this deception, and respond by rushing up and trying to catch the supposed prey. With some practice it's possible to lure the tarantula at least part of the way out of the burrow so it can be pinned to the ground with a fast but gentle slap of the hand, and captured. Individual tarantulas differ in how determined they are to catch the insect they think has stumbled down their burrow—some spiders will make a few half-hearted inquiries with their front legs and then ignore any

Figure 8.9 "Fishing" for tarantulas.

further temptation, while others are so stimulated by the vibrations that they actually run up the twig right into the collector's hand.

There is definitely some art involved in mastering this technique. Manipulating the twig in too clumsy a manner will only frighten the tarantula and drive it deeper down the burrow. The collector has to know just how much twig movement will keep the spider interested and bring it to the surface. Some collectors like to put a piece of sticky material, like chewing gum, on the end of the twig in the belief that the spider will have difficulty disengaging its fangs and can be forcibly pulled out of the burrow. Many children in Belize place the tip of the twig in their mouths before using it. They believe that tarantulas are attracted to the saliva. I have not found these modifications to provide any advantage. Tarantulas have to be coaxed or fooled out of their burrows. Practice is the only way to develop proficiency, and there will be many frustrating attempts—and lost spiders—on the way to expertise. Tarantulas will not respond to an incorrectly manipulated twig indefinitely, and after two or three failed attempts the spider will become wary of the bait and ignore further attempts to lure it outside.

Once lured to the surface, the tarantula must be restrained quickly and surely or it will retreat back into its burrow and be lost. Your flattened hand should be brought down rapidly over the spider, pinning it firmly against the ground. With its ventral surface pressed tight against the earth, the spider can't bite. At this point you can relax a little—the spider's not going anywhere—and decide the best way to pick it up for closer examination. The burrow entrance can be blocked by an object such as a stick or anything you may have brought along: a camera, backpack, canteen, cigarette pack, etc. Having made escape down the burrow impossible, the spider can be released from under your palm and coaxed into a container. Alternatively, those who are more comfortable around these spiders can pick them up by grasping them with thumb and forefinger on either side of the carapace, between legs II and III, and lift them up off the ground.

With extremely shy spiders, or if a nighttime visit isn't possible, tarantulas can simply be dug out of their burrows. Of course in doing this, the tarantula's home is destroyed. It is unethical to dig a tarantula out of its burrow unless the intent is to capture the spider permanently and there is a justification for doing so. Never dig up a burrow only to release the tarantula later—the spider will probably not survive. A large investment of energy is required to excavate a burrow, and a tarantula that is simply thrown back into the forest to fend for itself will have a poor chance of survival. Therefore the only time when it is appropriate to dig up a burrow is when the spider is going to be kept alive in captivity or humanely killed and preserved for scientific study.

Experience has taught that the best way to dig out a tarantula is to insert a finger down the burrow first. Then begin methodically removing the soil around the finger with a hand spade, inserting your finger ever farther down the tunnel to follow the twists and turns until the spider is encountered (Fig. 8.10). Most tarantulas will retreat to the end of the burrow during this process and offer no defense when met by your fingertip, but occasionally an individual will react to this invasion of its home by biting. For anyone who wants to avoid the possibility of being bitten, an

Figure 8.10 Finger insertion method.

alternative technique is to run a stiff wire as far as possible down the burrow and begin digging without inserting a finger. The wire serves as a guide for following the burrow if the surrounding soil collapses and fills the tunnel. It is important to bend the end of the wire into a blunt hairpin before it is inserted, otherwise the sharp tip could impale the tarantula. Using a finger to follow the burrow is the preferred method because it prevents injury to the tarantula and keeps the trail to the bottom from being lost by preventing loose soil from filling the tunnel.

## Finding Tarantulas in Trees

Arboreal tarantulas are much harder to find than burrow dwellers. Look for traces of their silk retreats on the sides of trees, especially in spots with loose and partially peeling bark. Another good place to look is along the edges of roofs, particularly the corrugated tin kind, because arboreal tarantulas often construct tube webs under such structures. At night, by searching with a flashlight along the overhang of roofs, these spiders can sometimes be seen with their foreparts outside the tube web waiting for prey. If their retreats are low enough to be reached, arboreal tarantulas can be lured out by the same fishing technique described previously. However, most of these spiders are living high in the canopy where they remain virtually inaccessible. The only time that a hunt for arboreal tarantulas offers much promise is during the breeding season, because the males will descend from the canopy and rest on tree trunks at night, low enough to be seen.

# Chapter 9

# Collecting and the Law

Persons who simply want to observe tarantulas in Belize, and perhaps briefly capture them to take a photograph and then release them, do not need to obtain permits unless they are doing so in a protected reserve like the Bladen Nature Preserve. Belize's designated conservation areas, like U.S. national parks, have rules which prohibit direct contact with wildlife. At other locations, it is always a good idea to find out if the land is privately owned, and if so, to courteously introduce yourself to the owners and obtain their permission before hunting for tarantulas. Poking around in the weeds, especially at night with a flashlight, can look very suspicious to people living nearby.

Anyone wishing to collect tarantulas and keep them in captivity for an extended period, or to transport them to another country, needs to follow certain procedures to comply with wildlife conservation laws. There are legitimate reasons for permanently removing tarantulas from the wild. Type specimens of new species or vouchers for significant new locality records must be deposited in museums to be available for re-examination by other scientists. Species described without representatives set aside for future reference do not help advance biological knowledge; they only cause confusion. A large amount of what is known about the behavior and reproductive biology of tarantulas was gained through the observations of scientifically oriented hobbyists. Collecting tarantulas in small and appropriate numbers with the intention of studying them and publishing these findings to the community of arachnologists, whether by professional or amateur students, is another justification for removing spiders from the wild. These activities are quite distinct from commercially motivated collecting or the desire to possess a trophy or souvenir, and should not be confused.

## Collecting Permits

Many people are surprised to learn that it is illegal to collect tarantulas in Belize (as in many places around the world) without permits. The Wildlife Protection Act of 1981 covers all Belizean animals including invertebrates. Anyone who wants to collect tarantulas and keep them in captivity must first apply for a scientific collecting permit. The Forest Department's Conservation Division is the permit granting authority. The wildlife officers will require a summary of the proposed study with a justification for collection, basic information about the applicant and their institutional affiliation if any, the locations where the collections will take place, and the species and numbers of specimens needed. A fee is charged for a scientific collecting permit, currently US$100.00, and it remains valid for 1 year.

Failure to follow the wildlife laws of Belize will result in the specimens being seized, and fines or imprisonment for the violator depending on the circumstances. Ignoring the law also shows a disregard for the efforts of the government to protect wildlife and discredits responsible biologists who work in honest partnership with conservation authorities. Permit application procedures are reasonable, and legitimate requests are given fair consideration, so there is no justification to ignore the rules.

## Export Permits

Before any tarantulas can be taken out of Belize, an export permit must be obtained. This requirement is in addition to a scientific collecting permit, which only covers the act of handling and capture within Belize. The Forest Department also is responsible for issuing these permits. A collector who possesses a scientific collecting permit will usually be required to bring the specimens to Forest Department headquarters in the capitol city Belmopan, where the spiders will be inspected prior to issuance of an export permit. Because the genus *Brachypelma* is listed under Appendix II of CITES, an export permit for Mexican redrump tarantulas, including dried specimens or preserved parts, must be on an official CITES permit form to avoid problems and perhaps seizure by law enforcement officials at the port of entry in the country the spiders are being exported to. For all other species, export permits can be printed on Forest Department letterhead with the appropriate signature and stamp. Export permits are issued for single events and are only valid for a short period. If successive exportations are planned, separate applications will have to be submitted for each shipment. No fee is charged for an export permit.

## Packing for Shipment

Live tarantulas being exported should be individually housed in plastic containers such as delicatessen cups or margarine tubs, with tight-fitting lids. Approximately a dozen air holes the width of a pencil lead should be punched in the side of the container, pushing through from the inside out so no sharp projections which could injure the tarantula are pointing inward. A soft and absorbent material like paper towel or toilet tissue needs to be arranged inside the container to create a hollow space just large enough to accommodate the spider in a crouching stance with packing material above, below, and on all sides (Fig. 9.1). The air holes must be large enough to prevent suffocation, but small enough to keep the humidity high within the cup. Never pack the spider in soil or leaves because in most countries, this will result in the shipment being confiscated upon its entry by agriculture inspectors. These agencies are charged with protecting agricultural interests within their country and are concerned that foreign soil or plant parts might introduce plant diseases or insect pests. Sprinkle some water onto the towels just before the lid is snapped closed to prevent the spider from desiccating during the trip. It is a good idea to secure the lid shut with some stout tape because large

Collecting and the Law 71

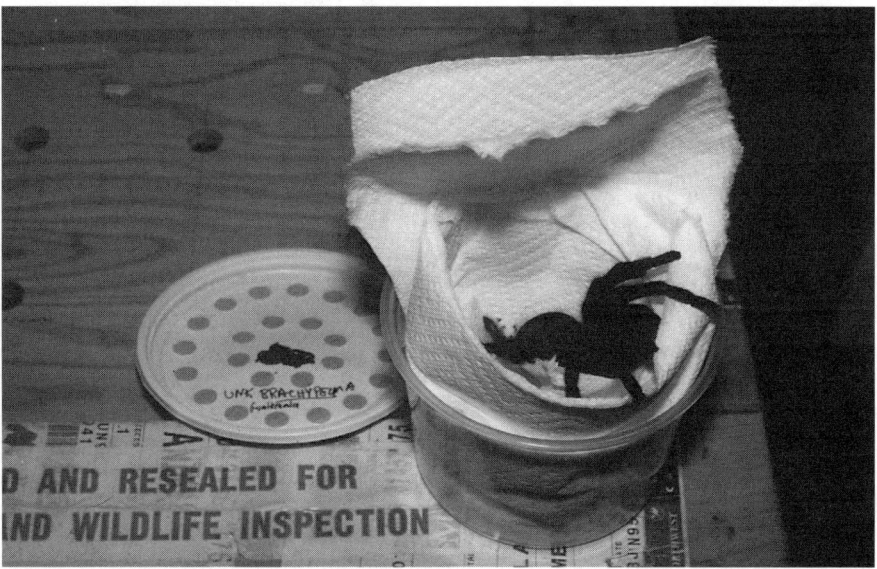

Figure 9.1 A tarantula packed for shipment.

tarantulas are strong and may be very persistent in their efforts to force open the top and escape the tight and unfamiliar surroundings, making subsequent entry inspection more exciting for wildlife agents than it ought to be.

## Other Regulations

Although scientific collecting and export permits make taking tarantulas out of the country legal, the laws of the country to which the specimens are being sent must also be complied with. As these regulations will vary it is recommended that the collector query the appropriate authorities in advance to learn what will be required. In the United States, all live or preserved tarantulas must be declared and made available for inspection at the time they enter the country. Only certain cities are designated as ports for lawful entry of foreign wildlife, so the specimens' itinerary needs to be routed so the first U.S. city they arrive in is a designated port. At least 24 hours prior to the time the specimens are due to arrive in the United States, the exporter should contact the law enforcement office of the USFWS in the port of entry and advise them of the estimated time of arrival, the name of the airline carrier, the species and numbers being shipped, and their condition (whether alive or preserved). The declaration form for wildlife differs from the one most travelers fill out to declare overseas purchases to U.S. Customs. Wildlife and all wildlife products are declared on the form 3–177 which is obtainable from any office of the USFWS. A copy of the completed form should be faxed to the USFWS office when the inspectors are notified of the impending shipment. You should arrange the layover between connecting flights beyond the port of entry city and your final destination to give you adequate time to clear your shipment—at least 2 hours is recommended—and if you are arriving into the port of entry city after 5 PM or on the weekend you may need to lay over a day or so in order to have the shipment inspected and cleared, or else be prepared to pay for the wildlife inspector's overtime expenses (if this has been prearranged).

Upon arrival at the U.S. port of entry you will inform the customs agent that you have wildlife to declare and that the USFWS agent has been notified. The customs agent will assist in summoning the wildlife inspector if one is not already present. The USFWS agent will inspect your spiders and examine your permits, and will keep the original export permit document (you should have a photocopy made for your records before leaving Belize, as you will have no opportunity to do so afterwards). If everything is in order and your specimens match what your permit authorizes you to export, the agent will stamp your 3–177 form "cleared" and you will have legally exported tarantulas into the United States.

Since wildlife regulations are subject to regular change, with the tendency to become ever more restrictive, it is a good idea to contact the Belize Forest Department and the enforcement agency in your home country before considering collecting any tarantula in Belize.

# Chapter 10

# How to Identify Belizean Tarantulas

In many parts of the world, determining what kind of tarantula is in hand can be a frustrating task. A large number of tarantula species are medium-size, earth brown spiders and are unremarkable and hard to differentiate. Identification of species, and sometimes genus, requires close inspection of minute structural details that can only be done in a lab equipped with a microscope, dissecting scope, and a stack of old manuscripts. Fortunately, most Belizean tarantulas can be distinguished by color, pattern, or obvious structural design.

## General Morphology

In certain cases a closer inspection is required to assure an accurate identification, and this calls for a familiarity with tarantula morphology (Figs. 10.1, 10.2). The tarantula body is divided into two parts, the cephalothorax (or prosoma) and abdomen (or opisthosoma). The top of the cephalothorax is covered with a hard plate called the carapace. On the underside of the cephalothorax is a central plate called the sternum, with the basal segments of the legs and the mouthparts radiating out from it. The abdomen is attached to the cephalothorax by the pedicle, a thin stalk through which run the central nervous chord, digestive tract and circulatory system. Visible on the underside of the abdomen are the four book lung openings and epigastric furrow (anterior) and the spinnerets and excretory opening (posterior).

Although tarantulas have eight legs, an examination by someone unfamiliar with these spiders will reveal what appears to be a fifth pair of short legs at the anterior of the cephalothorax. These appendages are the pedipalps, which are derived from the spider's mouthparts and are not true legs. Pedipalps are used to manipulate prey, grasp the eggsac, and are transformed in the mature male into mechanisms for inseminating females.

Between the pedipalps, at the front of the tarantula, are the chelicerae. At the end of the two chelicerae and normally kept folded tightly against them are the fangs. On the underside of the chelicerae are hard projections which create an opposing surface for the fangs to grind prey against during feeding.

The eight true legs are designated from front to back as leg I, II, III, and IV. Each leg is divided into seven segments. Moving outward from where they join the body, these segments are: coxa, trochanter, femur, patella, tibia, metatarsus, and tarsus. The pedipalps, betraying their different origin and function, lack the metatarsal segment.

At the end of every tarsus are two tiny claws which enable the tarantula to grasp

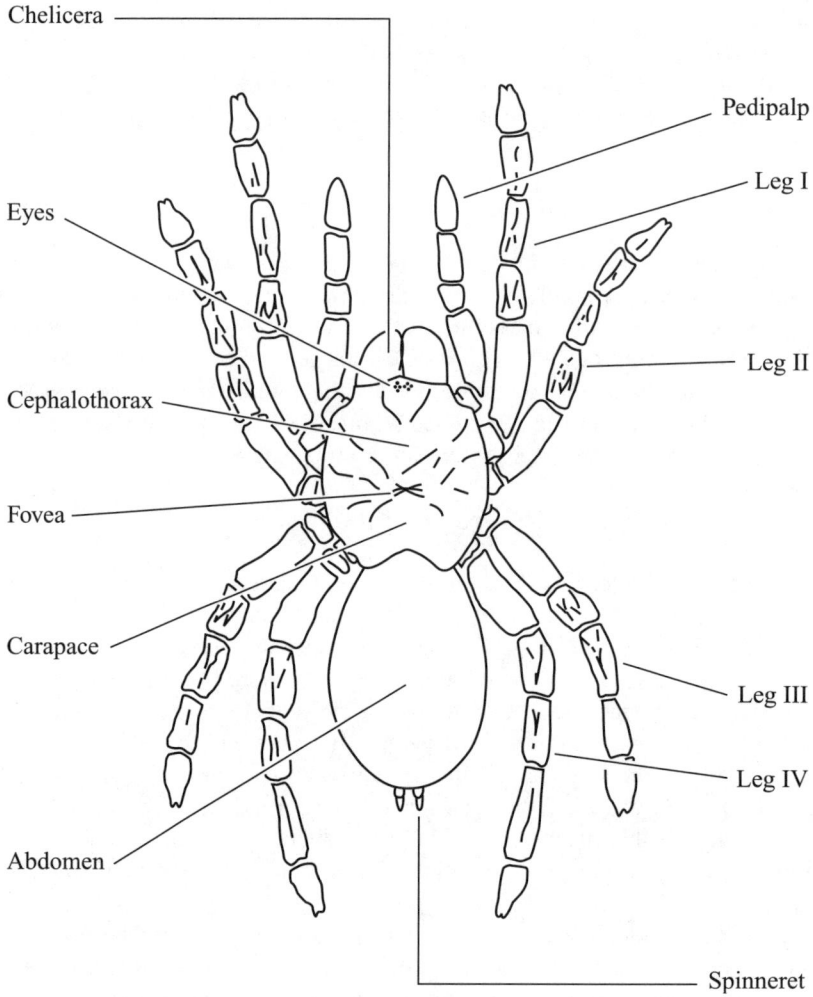

Figure 10.1 Dorsal morphology.

rough surfaces when climbing. Tarantulas also are able to scale very smooth surfaces and this is possible due to highly modified setae on the undersurface of the tarsi. Each tarsus and most metatarsi have many tiny scopula setae. Each seta is frayed at its end, creating a tremendous surface area and imparting an adhesive quality. The underside of each tarsus is completely covered with these setae to create a scopular pad, and it is with this structure that a tarantula can walk straight up a pane of glass.

Most tarantulas are quite similar in overall morphology and only a small percentage of species have body forms or structures that diverge from the basic design. In contrast, their setae exhibit myriad shapes and functions. The defensive

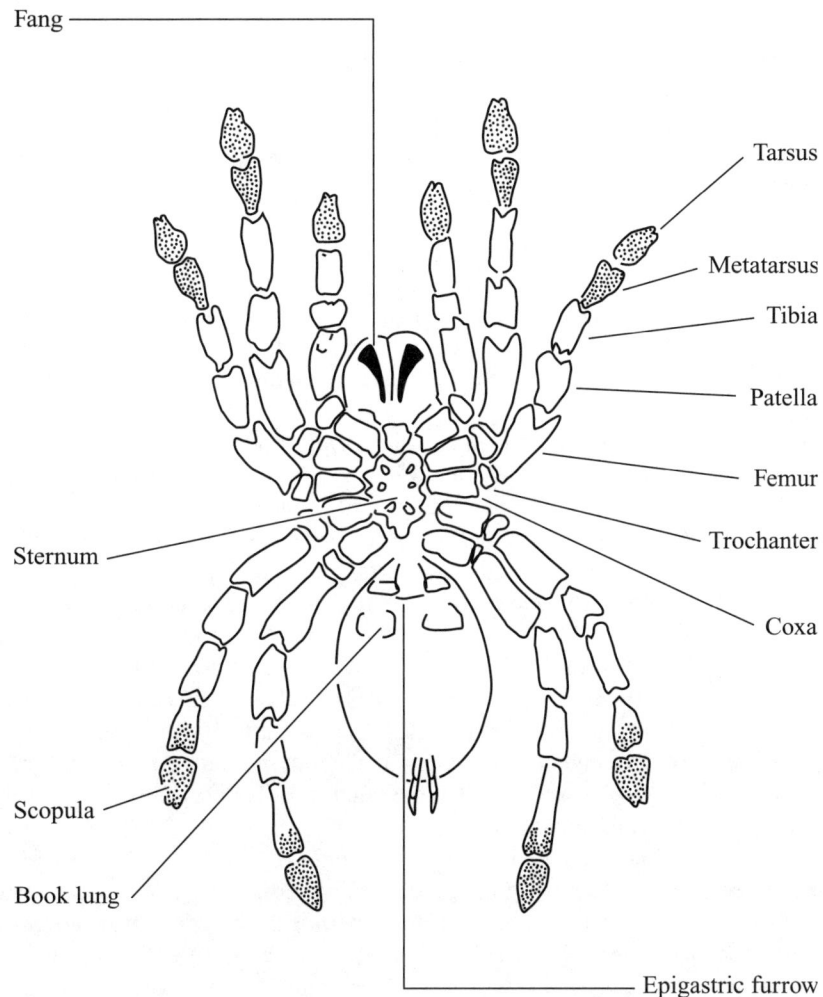

Figure 10.2 Ventral morphology.

use of urticaceous setae and the adhesive quality of scopular setae have been described. Some setae are clavate and detect air movement, thus serving as sensitive motion detectors. Other setae are plumose and may produce sounds when rubbed against an opposing body surface. No Belizean tarantula makes sounds audible to human ears, although it is possible that other animals can detect them. The colors and patterns of tarantulas, sometimes startlingly beautiful, are imparted by the setae. All Belizean tarantulas are clothed in a short undercoat of pubescence which is colored in somber earth tones. The legs, abdomen, and chelicerae have a second coat of longer hairlike setae that may be brightly colored. The setae are perhaps the least constrained aspect of the tarantula *bauplan,* and they are structures which express both the different evolutionary histories as well as the varied lifestyles among tarantulas. In contrast to many other animals where differences in gross morphology are apparent, determining relationships between tarantula species often rests heavily on comparative study of seta structure and distribution.

With this basic knowledge of tarantula morphology, one may look for variations in these features that identify each species.

## Diagnostic Characters

In the instances where more than a cursory glance at a tarantula is necessary for species identification, five features are most important: spinnerets, male embolus and tibial spur, female spermatheca, and the setae on the prolateral surface of the femur of leg I.

Spinnerets are fingerlike extensions at the posterior of the abdomen. Although two pairs are present, only the posterior pair are large enough to be easily visible. Each posterior spinneret is composed of three segments, and the shape of the apical segment will distinguish barychelid pygmy tarantulas from typical tarantulas. In pygmy tarantulas the apical segment is short and buttonlike, barely longer than wide, while in all other Belizean tarantulas it is long and fingerlike, and much longer than wide.

Male tarantulas become sexually mature suddenly, after the ultimate ecdysis of their exoskeleton. Up until that time they closely resemble females in superficial structure. Upon their final molt, mature males are equipped with two new structures. The first is a profound modification of the pedipalps. Each tarsus of the pedipalps, which previously was identical to the other tarsi, is transformed at maturity into the cymbium. At the end of the cymbium is the palpal bulb, which is used for depositing semen into the female. The hollow bulb tapers to a narrow embolus. Along the embolus runs a groove where the seminal fluid flows from the bulb into the female's spermatheca. The shape of the embolus is variable between species but highly conserved intraspecifically. As a result, embolus morphology has traditionally been the key character used by taxonomists when diagnosing new species and determining relationships between taxa. This practice has been convenient because the wandering males are the easiest tarantulas to find, especially during brief surveys of a region. The disadvantage is that a reliance on mature male characters for species diagnosis says nothing about how to identify females. Many species of tarantulas have been described solely on the basis of the mature male (including one by the author of this book), with no mention of female characters. As an illustration of the obstacles created by the sexual specificity of key taxonomic characters, take for example the population of pygmy tarantulas in the Cayo District. Many females have been found, but no males. The females possess several distinctions that provide strong evidence that they are not the same species as the pygmy tarantulas occurring in Orange Walk and Belize districts. However, it would be unwise to formally describe the Cayo spiders as a new species until a mature male is found, so that a wider array of reliably diagnostic characters can be established. Finding mature pygmy tarantula males will not be easy because of their small size, the best methods being long-term pitfall trap sampling or collecting gravid females and rearing the spiderlings, both very labor- and time-intensive undertakings. Another example of the pitfalls of species descriptions based on a small sample size is the question of the validity of the Yucatán rustrump tarantula, *B. epicureanum,* found in neighboring Mexican states in the Yucatán Peninsula. This species was described from two specimens in 1925. The differences between this spider and *B. vagans* are slight, and poorly delineated in the original paper. From the original description it is impossible to determine if there is any diagnostic character that would confirm the validity of *B. epicureanum.* Only by revisiting the type locality and examining additional specimens will it be possible to confirm that the redrumps in the northern Yucatán are a distinct species, and this has yet to be done. During the late 1800s and early 1900s some scientific papers naming new species were accompanied by such vague descriptions that they could have been applied to many previously described species. These papers are sometimes so uninformative that they offer few clues as to why the scientist felt that the animal warranted a new species designation. Most modern species descriptions are based on both male and female characters, and whenever possible, on multiple specimens of both sexes to gauge the variability of the characters. Arachnologist Tom Prentice has been most influential in introducing this practice to tarantula systematics by authoring several new species descriptions with a level of meticulous detail unmatched by any previous work.

Under magnification the embolus is revealed to be a fairly elaborate and intricately adorned structure rather than just the tiny spur it appears to be at first glance with the naked eye. Although the basic design is a broad base tapering toward the tip, there are many variations on this plan. Some emboli are broad and blunt while those of other species may be extremely thin and delicate (Fig. 10.3). The apex may curve upward and be somewhat twisted. In some species, sharp-edged keels or ridges are clues for identification. This difference between species is undoubtedly an isolating mechanism to prevent inbreeding—analogous to a key that will only fit a specific lock, the "locks" being the female spermathecae which exhibit an equally varied range of shapes.

The male embolus is probably the single most important structure for distinguishing tarantula species. A strong hand-held magnifying glass is usually sufficient for seeing enough detail to make an identification based on embolus structure. A basic shape distinguishes genera with minor details pointing to species. The distal segment of the emboli of *Brachypelma* is spoon-shaped, that of the *Crassicrus* is a six-keeled upturned spiral, and in *Psalmopoeus* it is a long needle-like projection, and so on. If two closely related and similar looking tarantulas, such as the Livingston and Cayo tarantulas, are viewed together they may not impress the observer with any obvious differences, but if the emboli are compared a clear distinction will be seen.

The second male structures useful in identification are the tibial spurs. Male tarantulas possess two, one, or no tibial spurs on each tibia of leg I, depending on genus. In all but one Belizean species the condition is two spurs of unequal size on each leg, and this similarity reduces their utility as a field identifier. The sin-

Figure 10.3 Comparative morphology of male emboli in four Belizean tarantulas (clockwise from top left): Cayo, cinnamon, Gutzke's, and pygmy tarantulas. (Illustration by Norma Reichling)

gle exception is Gutzke's tarantula, which does not develop tibial spurs after the ultimate male molt.

The spermathecae are an important diagnostic tool but have limited value in the field. To be seen, the specimen must either be killed and dissected or kept alive until it molts whereupon a good facsimile of the structure will be provided on the inner ventral surface of the opisthosomal exuvia. As with the embolus, broad categories of spermatheca shape are useful as generic distinguishing characters while slighter variations will help to discern species (Fig. 10.4). The spermathecae of *Brachypelma* and *Reichlingia* species are fused and those of *Citharacanthus, Crassicrus,* and *Psalmopoeus* species are divided into two separate receptacles.

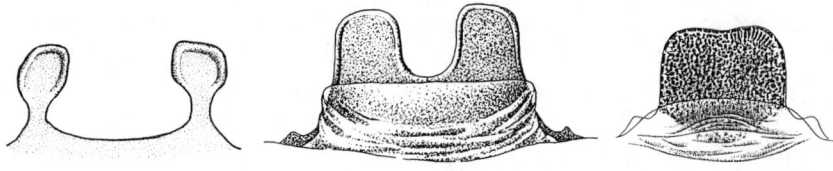

Figure 10.4 Comparative morphology of female spermathecae in three Belizean tarantulas (left to right): Cayo, cinnamon, and pygmy tarantulas. (Illustration by Norma Reichling)

The morphology of the spermatheca of Gutzke's tarantula is unknown since a female has never been found, but in putative congeners it is fused.

The gross structure of the setae on the prolateral surface of the femur of the first leg can be seen with the aid of a magnifying glass. Two broad categories can be described, filiform and plumose. Simple filiform setae lack ornamentation, barbs, etc., and resemble mammalian hair under low magnification. Plumose setae are so named because they are reminiscent of certain types of feathers such as ostrich plumes. These specialized setae have numerous side branches extending out in whorls from the main shaft. Plumose setae may be thin with only short branches (filiform plumose) or plump and capitate. These setae occur in dense patches on various parts of the body of some species, particularly at the bases of legs. Such patches are often called stridulation organs, although in many cases, e.g., *Citharacanthus* spp., it is not known for certain that the spider uses them to produce sound. Belizean species of the genera *Reichlingia, Crassicrus, Metriopelma,* and *Psalmopoeus* lack any plumose setae on the basoprolateral surface of femur I. *Brachypelma* species have filiform plumose setae on femur I. In *Citharacanthus* species these setae are pronounced on the femora and trochantera of legs I and II, and diagnostic both generically and specifically.

# Key to the Tarantulas of Belize

**ADULT MALES** (highland pygmy tarantula unknown):
1. No spurs on tibia I..................................................Gutzke's tarantula
   Tibia I with bifurcate spurs........................................................... 2

2. Apical segment of the posterior spinnerets short and
   buttonlike; all tarsal scopulae divided by a line of
   rigid, spinose setae; very small, legspan < 50 mm
   (2 in) (subfamily Trichopelmatinae).........................pygmy tarantula
   Apical segment of the posterior spinnerets long
   and fingerlike; tarsal scopulae entire; legspan
   > 100 mm (3.9 in)........................................................................ 3

3. Lyra present; scopulation on tarsi and metatarsi broader
   than tibiae; arboreal (subfamily Selenocosmiinae)...................... 7
   Scopulation on tarsi and metatarsi of equal or lesser
   width than tibiae; terrestrial (subfamily Theraphosinae)............... 4

4. Distal portion of palpal embolus with six
   prominent keels spiraling to a broadly truncated
   apex; tibia IV appears slightly incrassate..........................cinnamon tarantula
   Distal portion of palpal embolus with less than
   six or no keels; tibia IV not incrassate........................................ 5

5. Some abdominal setae bright red; a woolly
   tan fringe around the margin of the darker
   carapace; distal portion of embolus
   flattened and spathulate......................................... Mexican redrump tarantula
   Abdominal setae not bright red; no lighter
   woolly fringe on margin of carapace........................................... 6

6. Prolateral surface of trochantera I and II
   with scattered and capitate plumose setae
   amidst a field of very fine plumose setae;
   no prominent keels on embolus.......................................Livingston tarantula
   Prolateral surface of trochantera I and II with
   a dense field of very fine plumose setae,
   without capitate setae, apical division of
   embolus with five prominent keels...........................................Cayo tarantula

7. Color black................................................................ Maya tarantula
   Color brown or gray with long
   reddish setae on abdomen......................... Costa Rican orangemouth tarantula

**FEMALES** (Gutzke's tarantula unknown):
1. Apical segment of the posterior spinnerets
   short and buttonlike; dorsum of abdomen
   patterned with a series of light tan or
   apricot chevrons (subfamily Trichopelmatinae)........................................... 2
   Apical segment on the posterior spinnerets
   long and fingerlike; abdomen unpatterned................................................ 3

2. Double series of five light, irregularly shaped
   spots on abdomen; spermatheca fused and
   with smooth margins................................................................ pygmy tarantula
   Double series of six light chevrons
   on abdomen; spermatheca fused but
   anteriorly notched................................................ highland pygmy tarantula

3. Lyra present; scopulation on tarsi and
   metatarsi broader than tibiae; arboreal
   (subfamily Selenocosmiinae)................................................................ 7
   Scopulation on tarsi and metatarsi of
   equal or lesser width than tibiae; terrestrial
   (subfamily Theraphosinae)................................................................ 4

4. Tibia IV appears very incrassate........................................ cinnamon tarantula
   Tibia IV not incrassate................................................................ 5

5. Spermatheca fused................................................Mexican redrump tarantula
   Spermatheca bilobed................................................................ 6

6. Prolateral surface of trochantera I and II
   with scattered capitate plumose setae
   amidst a field of very fine plumose setae......................... Livingston tarantula
   Prolateral surface of trochantera I and II
   clothed in a dense field of very fine
   plumose setae, without capitate setae,
   on basoprolateral aspect........................................................ Cayo tarantula

7. Color black................................................................ Maya tarantula
   Color brown or gray with long
   reddish setae on abdomen......................... Costa Rican orangemouth tarantula

# Chapter 11

# Species Accounts

The following pages present a detailed profile of each species of tarantula occurring in Belize. The format and conventions used are outlined below.

### Description

This section is intended to supplement the photographs by emphasizing characters that make each species unique. Leg spans and body lengths are for mature spiders. Leg span measurements are from the apex of leg I to the apex of leg IV on the same side of the body. Body length is the distance between the anterior of the chelicerae to the posterior of the abdomen.

### Similar Species

To assist in cases where species identification is not certain, a description of distinguishing characters among similar appearing tarantulas is provided. If a sure identification is still not possible, consult the key presented elsewhere in this book.

### Distribution and Habitat

Trends in geographic distribution and correlation with habitat type are described and will provide some insight when studying the distribution maps. Some of the areas where species are said to occur are generalized due to insufficient collecting. Also described are the habitats and, when known, the microhabitat parameters that are relevant to each species.

### Maps

The maps display dots corresponding to specific sites where each species has actually been collected. Although, as pointed out earlier, tarantulas are rarely distributed over a large geographic area, the distribution of the dots on each map convey an accurate conception of the overall range of each species. The broken line corresponds to the 400-m contour line, and delineates the approximate boundary of the highland ecological zone. Specific locality data for every specimen examined is presented in the gazetteer.

## Abundance

The designation of a species as abundant or rare is based on my own collecting experience. Many tarantulas are abundant locally but may be rare over their distribution as a whole. Perhaps due to their reproductive strategy of producing many hundreds of offspring per eggsac, it is unusual for a tarantula species to be confirmed present in a specific location but rare. The more usual situation is for a species to be abundant in a specific location but absent in some adjacent areas that represent similar habitat.

## Reproductive Biology

The reproductive biology of most Belizean tarantulas is poorly known. Much of the information on reproduction for most species (worldwide) comes from the efforts of dedicated hobbyists who document and disseminate their observations on captives. In this section I have focused on data from wild individuals because aspects such as seasonality could be altered in captivity.

## Remarks

I have included notes on ecology and natural history whenever possible. For most species, there are still many gaps in our knowledge of these subjects. Also discussed are problems or controversies involving the taxonomy of the species.

# PYGMY TARANTULAS
## Family Barychelidae Simon, 1889
## Subfamily Trichopelmatinae Raven, 1985

**Belizean species** *Reichlingia annae, Reichlingia* sp.

The barychelids represent a distinctive group of "mini-tarantulas." The family is composed of numerous genera, exhibiting great diversity in Australasia. The neotropical species are less well known and there may be numerous undiscovered species. Belizean barychelids exhibit a division of the tarsal scopulae, especially on legs III and IV, composed of a row of stiff, spiny setae. Divided tarsal scopulae are found in other mygalomorph taxa, and it has recently been shown that juveniles of many species possess this character until they reach a larger size. However, in these species the setae forming the division are soft and hairlike, not the stiff spines seen in barychelids. The diminutive pygmy tarantulas are notable not only for their small size (Fig. 11.1), but because they lack urticaceous setae.

The barychelids resemble another group of small mygalomorphs, the theraphosid genus *Acanthopelma* (with a species occurring very near Belize), but with good magnification barychelids can be differentiated by the tarsal trichobothria being distributed across the entirety of the segment, rather than being limited to the distal 2/3, and by a weaker development of the maxillary apophysis. Spinneret morphology serves as a more convenient field character.

Figure 11.1 Pygmy tarantula on fingertip.

## Pygmy Tarantula *Reichlingia annae* (Reichling, 1997)

### Description
The apical segment of the posterior spinnerets is short and buttonlike, shaped like a dome or a helmet. Males and females differ significantly in color and pattern.

*Male* Carapace black with coppery margins. The dorsum of the abdomen is a solid copper color with four faint black smudges on each side. On the spider's underside, the sternum is black and the abdomen is the same coppery color as the dorsum. The legs are charcoal black with copper highlights on their bases.

The twisted embolus possesses a subapical apophysis.

The smallest tarantula in Belize, and one of the smallest species in the world. Male leg span 4.5–5 cm (1.8–2 in), body length 1.5–2 cm (0.6–0.8 in).

*Female* Carapace medium brown with pale golden highlights. The abdomen displays a distinctive pattern consisting of a double series of five irregular tan or drab orange spots (Plate 11.1). In some specimens these spots are so reduced as to be nearly absent, while in others they connect along the middorsal line to form five conspicuous chevrons. These markings are on a medium to dark brown background. The entire undersurface is brown. The spermatheca is fused and roughly the shape of a square, with a smooth margin.

Leg span 3–4 cm (1.2–1.6 in), body length 1.5–2 cm (0.6–0.8 in).

### Similar Species
This and the highland barychelid are the only Belizean tarantula species with a patterned abdomen. The Cayo spiders can be distinguished from the present species by their larger size, brighter abdominal markings, and notched spermathecae (as opposed to completely fused in *R. annae*). Occasional specimens of pygmy tarantulas exhibit chevrons rather than spots, but these are never as wide and bold as in the highland *Reichlingia* and never in six pairs as in the latter species.

This diminutive tarantula could be mistaken for a juvenile of one of the larger kinds. Pygmy tarantulas can be found living alongside Mexican redrump, cinnamon, and Cayo tarantulas. Juveniles of the larger species have unpatterned abdomens except for the patch of urticating hairs which creates a black spot on their rear.

*Reichlingia* species resemble a theraphosid species known from Guatemala, *Acanthopelma rufescens*. These taxa may be sympatric in western Belize (see comments in Chapter 12, Extralimital Species). Specimens of *Acanthopelma* will be distinguishable in the field by their fingerlike apical segment of the posterior spinneret.

Barychelid pygmy tarantulas might also be mistaken for one of a variety of other small, burrowing spiders. Lycosids are hirsute, live in burrows, and have patterned abdomens, but can be distinguished by having opposing chelicerae, conspicuous eyes, and spindly, tapering legs. A small black mygalomorph spider is sympatric with the pygmy tarantula, and its burrows are very similar. However, the silken burrow lining of the black mygalomorph is much looser and can be easily destroyed with a small twig, whereas the lining of a pygmy tarantula burrow is tougher and will not come apart easily.

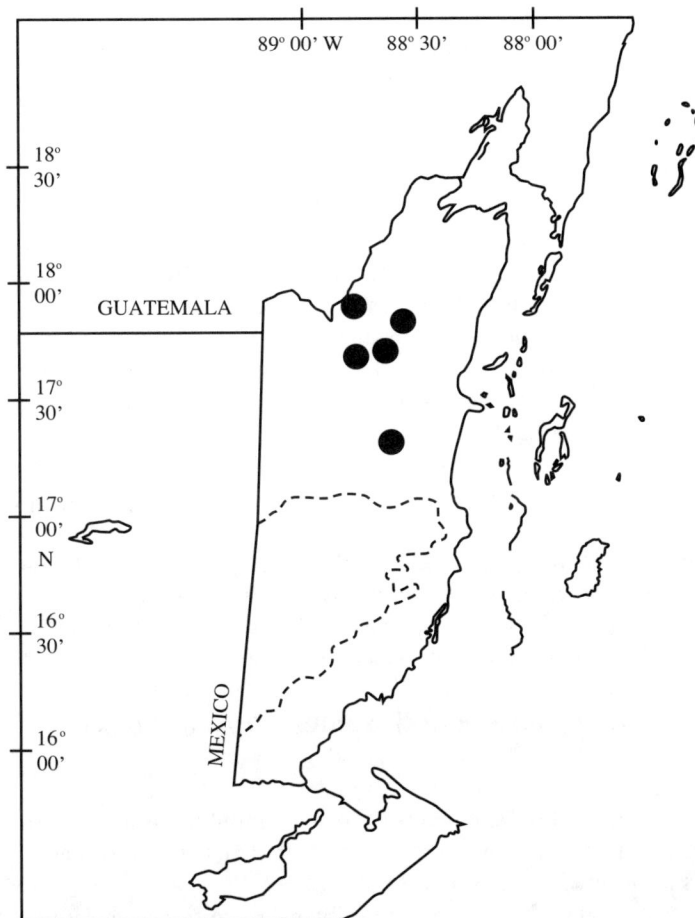

Figure 11.2 Distribution of the pygmy tarantula.

## Distribution and Habitat

The pygmy tarantula occurs in disjunct patches of pine savanna in Orange Walk and Belize Districts (Fig. 11.2). No specimens are known from Mexico or Guatemala. Substrate in this habitat is almost entirely sand to the depth of these spiders' burrows. The species is also found in forest patches that intervene the savannas, but is limited there to exposed sunny locations such as milpas and along foot trails. Elevation in both these habitats is less than 100 m (328 ft).

## Abundance

This species is abundant in savanna and pine ridge habitats, and less common where the substrate is not primarily sand. The pygmy tarantula is rarely found in undisturbed forest.

## Reproductive Biology

Females collected while gravid subsequently laid eggs in June, September, and October, which departs from the seasonal pattern of other tarantulas in the region. Eggs incubated in the lab at 23°C (73.4°F) produced emerging spiderlings after 35–37 days. One eggsac contained just 32 eggs, indicating a lower fecundity for this tarantula than most other species.

## Remarks

The burrow of this tiny spider is unique among Belizean tarantulas due to the presence of a side tunnel which usually forks off the main burrow. This extra passageway probably serves as an escape route in the event of invasion by a predator. If a burrow is severely disturbed the resident spider will sometimes appear unexpectedly at the surface several inches from the main burrow entrance. The pygmy tarantula is one of two species in Belize known to plug their burrow entrances (the other is the Mexican redrump), and burrow aggregations sometimes disappear from sight for months at a time only to reappear later in the year.

The pygmy tarantula was originally described as a member of the theraphosid genus *Acanthopelma*. A recent study revealed that it was incorrectly assigned to that family and was therefore transferred into the newly erected barychelid genus *Reichlingia*. The specific epithet is in honor of my wife, Ann.

## Highland Pygmy Tarantula *Reichlingia* species

### Description

Apical segment of the spinnerets short and buttonlike, as in *R. annae*. The male is unknown but presumably differs in coloration and extent of pattern, as in the case of the closely related *R. annae*.

*Female* A slightly larger and more intensely colored pygmy tarantula. Abdomen marked with a distinct series of six light chevrons on a dark background. Chevrons are apricot or burnt orange in color, and set off sharply against a much darker ground color which may be nearly black. Carapace and legs medium brown, as in *R. annae*. (Plate 11.2). The spermatheca is fused into a single square-shaped flap, but a distinct notch is visible in the middle of the anterior edge.

Attains a slightly larger size than *R. annae*. Leg span to 4.8 cm (1.9 in), body length to 2.6 cm (1 in).

### Similar Species

See comments under *R. annae* account.

### Distribution and Habitat

This population of pygmy tarantula appears to be geographically isolated from its close relatives in Orange Walk and Belize Districts. It occurs in the forested hills along the Macal River in western Belize (Fig. 11.3). Specimens have been

# Species Accounts

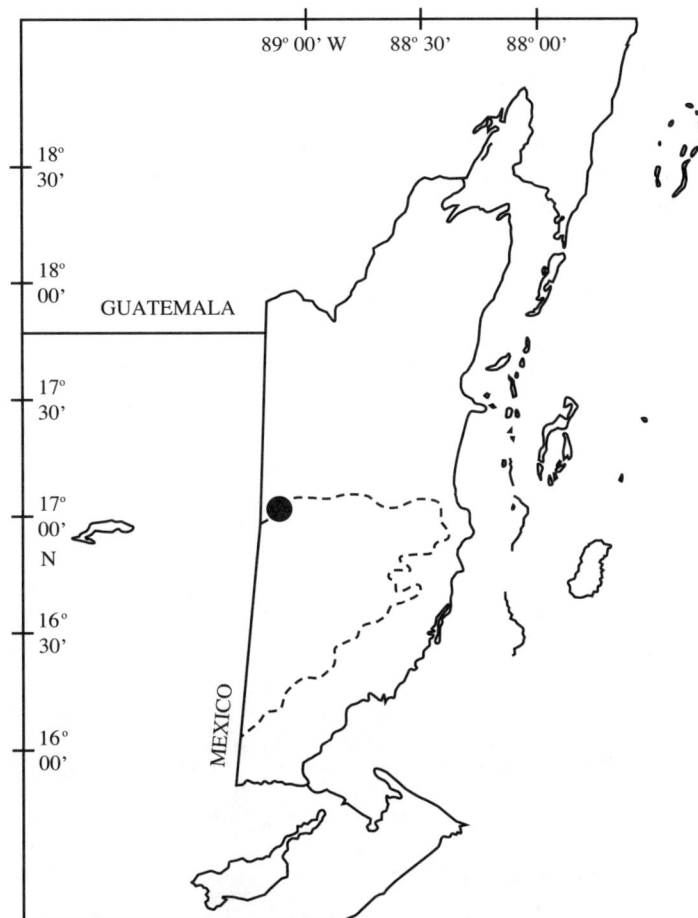

Figure 11.3 Distribution of the highland pygmy tarantula in Belize.

collected on shady, forested hillsides at elevations between 200–400 m (656–1311 ft). In contrast to *R. annae*, the pygmy tarantulas of the Cayo District are inhabitants of forested areas and avoid clearings.

## Abundance

The small size of the burrow opening coupled with the leaf litter and poor light conditions in the forest make this spider difficult to find. Within its limited distribution it is probably abundant.

## Reproductive Biology

Presumably differs little from *R. annae*, but no direct observations on this species have been made.

**Remarks**

To date only females have been found, making species determination uncertain.

The distribution of this spider requires further study. As currently known the area of occurrence is extremely small, and efforts to find it in similar habitat farther into the Chiquibul and Vaca Plateaus have so far failed. The first specimen discovered was extremely wary and could not be lured from its burrow, so an attempt to dig it out was undertaken. Unfortunately, the hunt was aborted after the tunnel was lost due to collapsing soil. A return to the site 6 months later found the burrow repaired, and a second attempt to excavate the specimen was successful.

Both forms of pygmy tarantulas are somewhat more irritable than other Belizean tarantulas. If held in the hand they will advertise this disposition by spreading their chelicerae and erecting their fangs, and may bite without further provocation.

# TYPICAL TARANTULAS
## Family Theraphosidae Thorell, 1869
## Subfamily Theraphosinae Thorell, 1870

**Belizean Species** *Brachypelma vagans, Citharacanthus livingstoni, Citharacanthus meermani, Crassicrus lamanai, Metriopelma gutzkei*

A hallmark of this subfamily is the variety of types and usage of urticaceous setae. Theraphosines possess combinations of type I, III, IV, and VI setae. The willingness of theraphosine tarantulas to use these defensive setae is revealed by the "bald" area on the abdomen of many individuals. Perhaps such spiders have had a recent encounter with a large predator such as a coati and had to shed their entire armament of weaponry to save their lives. If the spider can avoid a similar confrontation until its next molt, it will emerge from the old shell fully armed with a new coat of itchy barbs.

Although not all species are large, those found in Belize best fit the popular notion of what a tarantula is: large, stocky, hirsute, and imposing. Belizean theraphosines are terrestrial, but roving males and subadults of either sex are sometimes found crawling up the walls of buildings.

### Mexican Redrump Tarantula *Brachypelma vagans* (Ausserer, 1875)

**Description**

Freshly molted adults have velvety black legs and chelicerae. The carapace is also black with tan fringe around the margin. The hallmark of this species is its bright red abdomen—the "red rump"—which results from long crimson hairs overlaying a black velvety undercoat (Plate 11.3). As the months after molting progress, these distinctive colors become faded. Such specimens have dark brown legs and carapace, and the abdomen is muted brick red. Because the black pubescence is exposed when the abdomen distends, individuals which are well fed or just about to molt have less colorful abdomens, and may appear to be unicolor black spiders (Plate 11.4). Adults of both sexes are similarly colored, but the mature males are more lightly built, with a smaller abdomen and longer, more spindly legs (Plate 11.5). Newly emerged spiderlings are gray with a black spot of urticating hairs on their abdomen. Older juveniles are dark brown at first glance and must be closely inspected to discern the red abdominal hairs.

The embolus exhibits prolateral superior, prolateral inferior, and apical keels. Leg span up to 17 cm (6.7 in), body length to 9 cm (3.5 in).

**Similar Species**

Clean specimens are unmistakable, but identification can be difficult with spiders that have not molted for a long time. Faded specimens of the Mexican redrump tarantula resemble Livingston or Cayo tarantulas, and identification is complicated by the fact that Mexican redrumps partially overlap the distributions of both these species. Even severely faded specimens exhibit a woolly tan fringe along the margin of the carapace, which is always lacking in the other two species. Other less clear-cut characters which may help to distinguish these species are the thinner legs and sparser carapace pubescence on Livingston and Cayo tarantulas.

The Mexican redrump tarantula is the only member of the genus that has been found in Belize. Although the type locality of the Guatemalan species *B. sabulosum* is recorded at the Tikal archaeological site 25 km from the border, this species does not penetrate into the Maya Mountains (see comments in Chapter 12, Extralimital Species). There is also no truth to the rumor of a solid black species of *Brachypelma* living in Belize. Occasional specimens of an undescribed *Brachypelma* species have been offered for sale by hobbyists and commercial dealers. These spiders have been labeled "Belize black tarantulas," or stated to have been collected in Belize. No such spider is found anywhere in the country, so these spiders are either captive-bred hybrids (a practice not uncommon among hobbyist breeders of tarantulas) or collected elsewhere (probably somewhere in the Mexican Yucatán).

### Distribution and Habitat

Ubiquitous throughout Belize (Fig. 11.4). An adaptable species which is frequently encountered at high density in cleared pastures, residential backyards, garbage dumps, and early successional scrub as well as old growth forest. The species also is found on many of the "dry" cays offshore in small strips of strand forest, as well as Ambergris Cay (actually a part of the mainland which was cut off by a channel dug by the Maya). Many of these offshore populations are probably the result of accidental introduction by man. The Mexican redrump tarantula has not been found in the Mountain Pine Ridge, where the Cayo tarantula fills its niche as the large, burrowing tarantula species.

### Abundance

The Mexican redrump tarantula is the most abundant tarantula in the Yucatán. Burrows are typically found in dense aggregations, particularly in cleared habitat. The average distance between neighboring burrows in a milpa near Indian Church Village in Orange Walk District was 4.7 m (15.4 ft), with many smaller assemblages of 3–6 burrows situated in 4 m$^2$ (43 ft$^2$) plots. The species is less common in habitat where the forest canopy restricts direct sunlight from reaching the ground, although single individuals or small clusters of burrows can usually be located after a careful search.

### Reproductive Biology

The reproductive biology of this common tarantula is the most fully documented of any Yucatecan species. With the exception of pygmy tarantulas, the other Belizean species appear to follow seasonal reproductive patterns which are similar. Mature males begin appearing in August and are abundant during September and October. By the end of December only a few stragglers and late emerg-

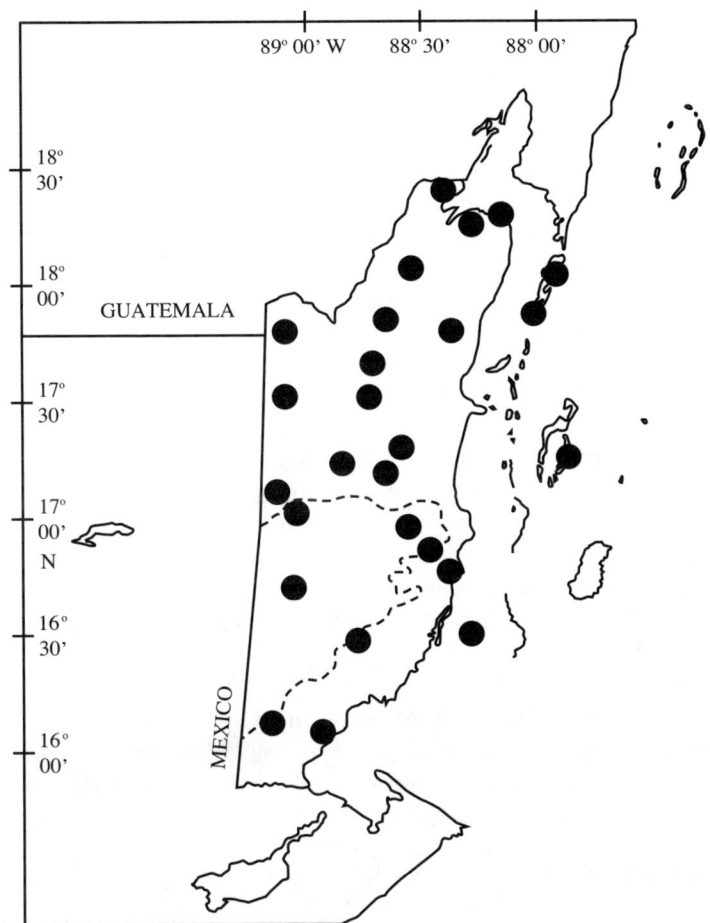

Figure 11.4 Distribution of the Mexican redrump tarantula in Belize.

ers can be found. The males are most active at night but also may be found roaming during the morning and late afternoon, or at any time during overcast days. Adult females molt once per year, just prior to the onset of the annual male emergence. Females collected in February and March have abdomens visibly distended with eggs, being large and round rather than more slender and slightly oval shaped. Most females lay their eggs in April, but in southern Belize some females have made cocoons by the middle of February. Most adult females observed during April and May are tending cocoons. A typical eggsac contains approximately 600 eggs. Spiderlings emerge from the cocoon after approximately 8 weeks, dispersing in aggregations during late May. By the middle of June all of the juveniles have left their maternal burrow, leaving the females to begin preparing themselves for another reproductive season.

### Remarks

Juveniles are commonly found inside rotting logs or hiding under surface debris. I have never found a Mexican redrump with a leg span less than 3.5 cm (1.4 in) living in a burrow of its own construction. In contrast, spiderlings will often construct burrows in captivity if maintained on a deep, moist substrate. Perhaps the greater effort required to excavate a burrow into the harder ground of their natural habitat prevents the attempt until the spiders have gained some size.

At sites where "redrumps" and cinnamon tarantulas occur together, individual Mexican redrump burrows are on average more distant from conspecifics than from burrows containing *Crassicrus*. This may indicate intraspecies competition, perhaps for prey, among cohorts.

The specific epithet is from the Latin *vagus*, "wandering," and probably refers to the behavior of mature males.

## Livingston Tarantula *Citharacanthus livingstoni*
## Schmidt and Weinmann, 1996

### Description

Males and females differ significantly in color.

*Male* Dark brown without markings, becoming black immediately after the ultimate molt (Plate 11.6).

The embolus is without prominent keels.

Leg span 12–16 cm (4.7–6.3 in), body length 3–4 cm (1.2–1.6 in).

*Female* The carapace is dark brown with very short, sparse pubescence. The chelicerae are gray. Dorsal surface of the abdomen is covered in a dense brown undercoat with a thick overlying layer of long, dull red hairs (Plate 11.7). The undersurfaces of the body and legs are shades of dark brown.

Leg span to 11 cm (4.3 in), body length to 4.6 cm (1.8 in).

### Similar Species

Very closely resembles the Cayo tarantula in gross morphology and color. Identification depends on an examination of the setae on the prolateral face of the trochanter of leg I and to a lesser extent, leg II. In *C. livingstoni* this area is covered with plump, capitate plumose setae scattered within a field of much finer setae. In contrast, the prolateral faces of these surfaces in *C. meermani* are clothed in a solid field of fine, plumose setae lacking thicker extensions. Mature male Livingston tarantulas are further distinguished from their Yucatecan congener by the simple unkeeled emboli.

### Distribution and Habitat

This species occurs in southwestern Belize and onward into the extreme southeastern corner of the Guatemalan Petén along the windward foothills of the Maya Mountains to the central ridge (Fig. 11.5). A *Citharacanthus* photographed near the summit of Doyle's Delight, the highest point in Belize at an elevation of 1124 m (3685 ft), by Sharon Matola, appears to be assignable to

Species Accounts 95

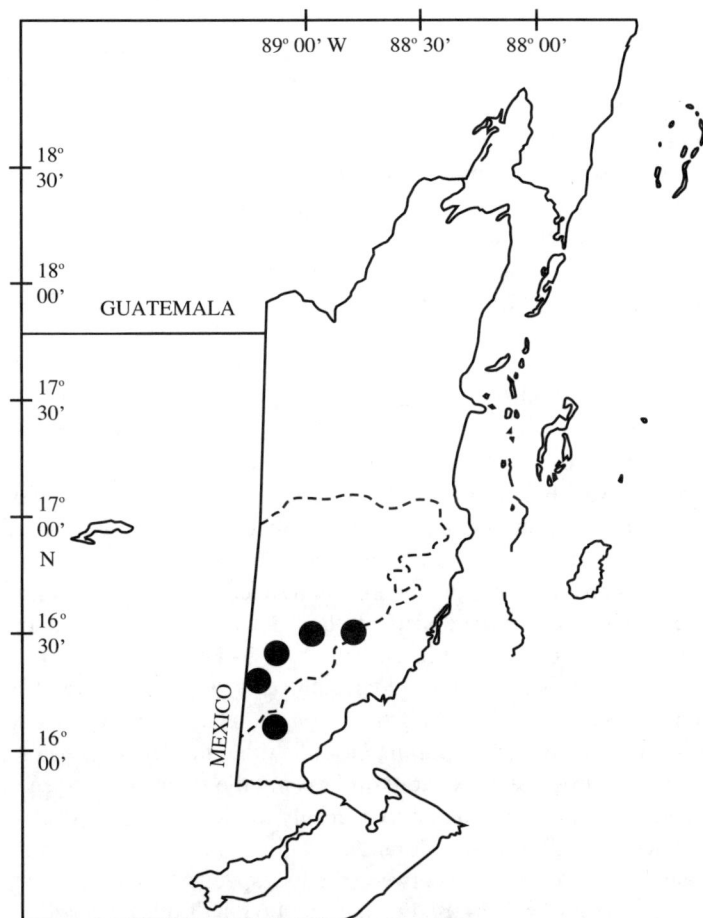

Figure 11.5 Distribution of the Livingston tarantula in Belize.

this species. The area intervening between the type locality of Livingston, Guatemala, and collection sites in Belize is a swampy lowland containing limited habitat suitable for tarantulas, and thus *C. livingstoni* is unlikely to be present in extreme southeastern Belize.

**Abundance**

This is an uncommon spider with most of its distribution in remote areas. Burrow aggregations have not been found.

**Reproductive Biology**

Nothing is known other than the holotype male was mature when collected on 13 September.

### Remarks

The Livingston, Mexican redrump, and cinnamon tarantula are sympatric within a narrow altitudinal range along the foothills of the Maya Mountains, between 200–400 m (656–1311 ft). All three species co-occur within this zone in the Bladen River Nature Preserve.

A nervous and fast-moving spider when outside its burrow. Specimens are difficult to hold in the hand without restraint due to their skittishness.

The specific epithet refers to the type locality.

## Cayo Tarantula *Citharacanthus meermani* Reichling and West, 2000

### Description

Males and females differ significantly in color.

*Male* The carapace is clothed in a woolly pubescence with a metallic copper sheen. The chelicerae are also coppery. The abdomen is black with long peach-colored hairs, sparsely distributed. The underside of the spider is gray on the sternum and black on the abdomen. The legs are metallic copper at their base (coxae and trochantera) and brown elsewhere (Plate 11.8).

The embolus possesses prominent prolateral superior, prolateral inferior, and subapical keels, and a weakly developed retrolateral and retrolateral accessory keel.

Leg span 12–16 cm (4.7–6.3 in), body length 3–4 cm (1.2–1.6 in)

*Female* A dull grayish brown spider with rusty abdominal setae (Plate 11.9). Not differing appreciably in color from Livingston tarantulas when spiders at the same point between molts are compared. Like other tarantulas, the colors on this species become progressively more dull during the months between molts. If a faded Cayo tarantula is compared to a freshly molted Livingston tarantula, they may be noticeably different in coloration but this is an artificial difference and not an accurate way to distinguish between the two species.

Leg span to 11 cm (4.3 in), body length to 4.6 cm (1.8 in).

### Similar Species

Although strong magnification is necessary to discern the plumose setae on trochanter and femur I, their effect of creating a sheen on these sections of the leg can be seen with the unaided eye. This effect is helpful in distinguishing Cayo tarantulas from faded Mexican redrumps, which also have dull red abdominal setae. When the surfaces on the legs are compared between the two species, the Mexican redrump's will appear dull and undistinguished as opposed to the velvety shimmer on *Citharacanthus,* which changes character as the leg is turned due to the changing angle of reflected light. See also comments under *B. vagans* and *C. livingstoni.*

### Distribution and Habitat

The Cayo tarantula is distributed across the Vaca Plateau in west central Belize and the eastern Petén of Guatemala (Fig. 11.6). It appears to be the only theraphosid spider living in the Mountain Pine Ridge. As with *C. livingstoni,* this species is primarily a forest dweller, preferring to establish its burrow in shaded

# Species Accounts

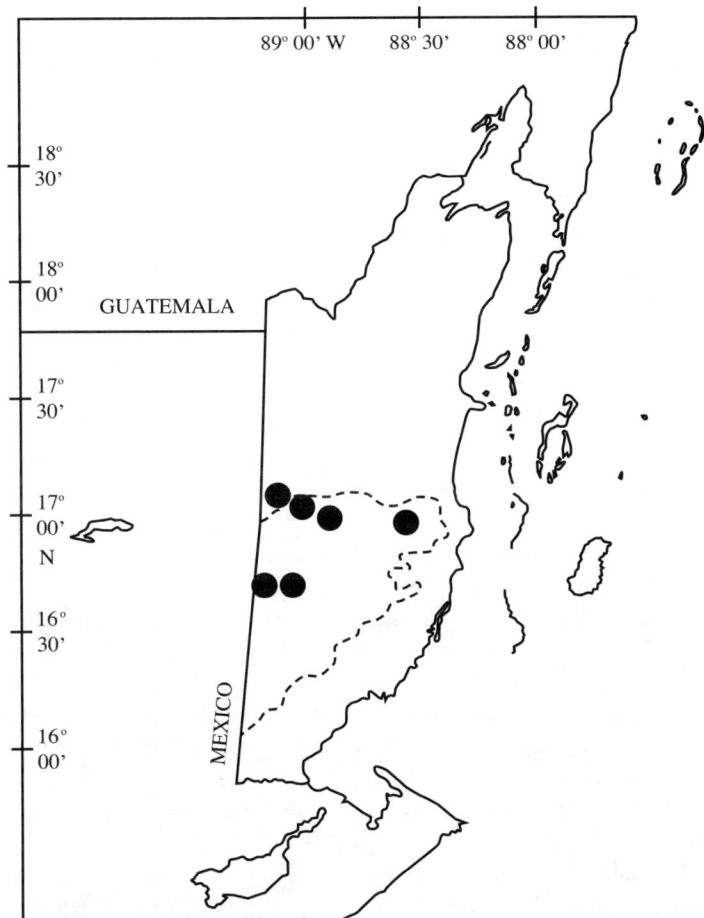

Figure 11.6 Distribution of the Cayo tarantula in Belize.

sites rather than in large open clearings as Mexican redrump and cinnamon tarantulas typically do. However, Cayo tarantulas do sometimes occur in small aggregations in clearings around residential yards, trails, or tree falls. In the Mountain Pine Ridge, *C. meermani* occurs in both the pine forest and the broadleaf forested canyon bottoms that transect the region.

## Abundance

The Cayo tarantula is abundant in suitable habitat. Unlike *C. livingstoni*, this species has been found in aggregations. Nine individuals were located within 1 hectare on a southwesterly facing slope in the Mountain Pine Ridge, and the type series included specimens collected from an aggregation along the Macal River.

## Reproductive Biology

One eggsac laid in captivity on 1 March contained 517 eggs.

## Remarks

This species is sympatric with Mexican redrump, cinnamon, and highland pygmy tarantulas in the hills southeast of San Ignacio, between 200–400 m (656–1311 ft) elevation. They are usually not far from forested areas and, unlike other sympatric species, avoid large expanses of open ground such as farms and cattle pastures. Several aggregations have been discovered alongside the mounds made by leaf-cutter ants (*Atta* sp.) in deep forest where the soft humus and open canopy created by the ants may provide an ideal microhabitat.

The specific epithet is a patronym in honor of Jan Meerman, a Dutch naturalist living in Belize, who was the first person known to collect specimens of this spider.

## Cinnamon Tarantula *Crassicrus lamanai* Reichling and West, 1996

## Description

Males and females differ significantly in color.

*Male* Mature males are entirely black with iridescent violet highlights visible under bright light (Plate 11.10). The lighter, leggy build characteristic of all mature male tarantulas is particularly evident in this species. Until the final molt males are colored similarly to females. The swollen tibia of the last leg, which is a hallmark of the female, is present but poorly developed in both immature and mature males.

The embolus angles abruptly. Six keels are present including prolateral superior, prolateral inferior, apical (serrated), and retrolateral.

Leg span 13–15 cm (5.1–5.9 in), body length 3.5–4 cm (1.4–1.6 in).

*Female* The overall appearance is of a brown spider, but the rich and varied tones on a freshly molted individual create an animal of subtle beauty. The anterior half is medium brown and distinctly lighter than the posterior half (Plate 11.11). Chelicerae, pedipalps, carapace, and first two legs are tan to light brown. Abdomen and rear legs are dark brown to nearly black. Long setae on the abdomen are unusually sparse, leaving the underlying pubescence more visible than on other Belizean tarantulas and giving the abdomen a distinctive velvety appearance.

The markedly incrassate fourth tibia develops as the spider matures and is only weakly inflated in young specimens.

Leg span 11–14 cm (4.3–5.5 in), body length 4–5 cm (1.6–2 in).

## Similar Species

*Crassicrus* is unlikely to be confused with any other species in Belize. The coloration of males and the incrassate tibia of females are diagnostic.

## Distribution and Habitat

This species presents a peculiar distribution pattern as it is absent from several areas which seem to represent suitable habitat (Fig. 11.7). *Crassicrus* occurs in the northern lowlands between Orange Walk Town and the Mountain Pine Ridge

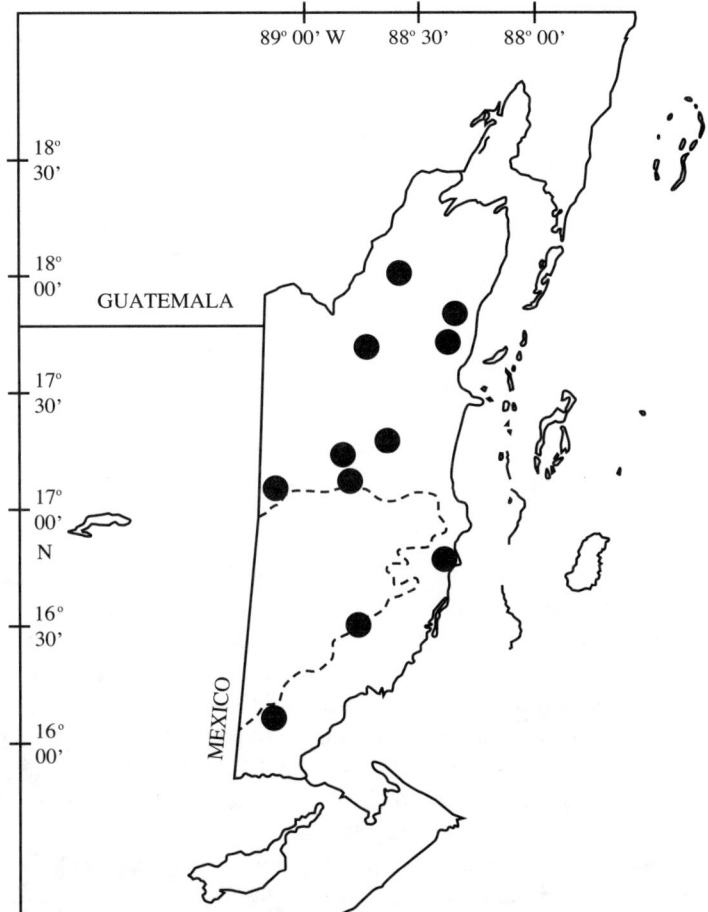

Figure 11.7 Distribution of the cinnamon tarantula in Belize.

and Vaca Plateau foothills. It can also be found on the coastal plateau bordering the Maya Mountains in southeastern Belize as far south as the Moho River. But strangely, cinnamon tarantulas have not been found on the Rio Bravo Escarpment, where it would be expected to occur based on habitat type and the presence of the Mexican redrump tarantula, with which it is completely sympatric. Specimens have been found outside the border of Belize, in Mexico as far north as Uxmal, Yucatán State (F. Vol, pers. comm., 2001), and the species is likely to occur in eastern Guatemala.

**Abundance**

The cinnamon tarantula is locally very abundant in Orange Walk District, typically occurring in aggregations of 40 or more closely spaced burrows. Occurrence is more sporadic in Stann Creek and Toledo Districts, where specimens are usually found as scattered, single individuals or in smaller clusters than is typical farther north.

**Reproductive Biology**

Females incorporate urticaceous setae shed from their abdomen into their cocoons—the only Belizean species that does so. Not only do the setae probably protect the eggs and spiderlings from predaceous flies as has been suggested for some South American species with similar habits, but they also impart a cryptic brown coloration to the otherwise bright white silk and make the cocoon (and thus the female as well) much less conspicuous when the tarantula carries it near the surface. By protecting her cocoon in this way, the female makes a trade-off because she loses most of her urticating hairs in the process and eliminates her primary defense against mammalian predators until her next molt. This species is very fecund considering its modest size; one eggsac was opened and found to contain 735 spiderlings.

**Remarks**

A local name for this species is antelope spider (R. West, pers. comm., 1995), referring to the mistaken belief that these tarantulas can leap great distances. This is based on the assumption that the swollen rear tibias are a muscular adaptation for jumping, like a grasshopper. It is true that sometimes when being handled a tarantula may make a short hop from hand to hand, but no tarantula can leap more than a few inches and terrestrial species rarely do so because their heavy bodies are fragile and can easily be damaged if they land on a hard surface. The true purpose of the thickened rear legs of this tarantula or the other species throughout the world exhibiting a similar condition is not known.

The specific epithet recognizes the pivotal role of the Lamanai Outpost Lodge in my study of Belizean tarantulas.

## Gutzke's Tarantula *Metriopelma gutzkei* (Reichling, 1997)

**Description**

The most beautiful tarantula of Belize. Males and females are likely to be different in appearance.

*Male* The carapace and legs are light grayish brown over which a metallic gold iridescence shines. Underlying abdominal pubescence is unusually sparse; most abdominal setae are of the longer, brightly colored type. The abdomen is clothed in fiery orange hairs, the color of paprika (Plate 11.12).

The embolus exhibits two keels, one subapical and the other prolateral inferior.

Leg span 10.5 cm (4.1 in), body length 3 cm (1.2 in). These measurements of the single known specimen, if representative, indicate that this is a small species.

*Female* The female of this species has yet to be found. Because the genus *Metriopelma* is notable for the number of species sporting bold abdominal patterns, and among patterned species generally male tarantulas are usually more uniformly col-

# Plates

Cover photo: Mexican Redrump tarantula. (Photo credit www.jackkenner.com)

Plate 11.1 Pygmy tarantula, female, Lamanai Archaeological Reserve, Orange Walk District. (Photo by Sam Fried)

Plate 11.2 Highland pygmy tarantula, female, San Ignacio Town, Cayo District.

Plates

Plate 11.3 Mexican Redrump tarantula, female, Lamanai Archaeological Reserve, Orange Walk District.

Plate 11.4 Mexican Redrump tarantula, female just prior to molting, Lamanai Archaeological Reserve, Orange Walk District.

Plates

Plate 11.5 Mexican Redrump tarantula, mature male, Indian Church Village, Orange Walk District.

Plate 11.6 Livingston tarantula, mature male holotype, Livingston, Guatemala.

Plates

Plate 11.7 Livingston tarantula, female, Bladen Nature Preserve, Toledo District. (Photo by Carrie L. Strehlau)

Plate 11.8 Cayo tarantula, mature male holotype, Cristo Rey, Cayo District.

Plate 11.9 Cayo tarantula, female paratype, Cristo Rey, Cayo District.

Plate 11.10 Cinnamon tarantula, male, Indian Church Village, Orange Walk District.

Plates

Plate 11.11 Cinnamon tarantula female, Indian Church Village, Orange Walk District.

Plate 11.12 Gutzke's tarantula, mature male holotype (preserved), Lamanai Outpost Lodge, Orange Walk District.

Plate 11.13 Maya tarantula, female, Sierra de Santa Cruz, Isabel, Guatemala. (Photo by Rick West)

Plate 11.14 Costa Rican orangemouth tarantula, subadult male, Blue Creek Rainforest Lodge, Toledo District.

ored than females (pygmy tarantulas are a good example), it is exciting to speculate that the female Gutzke's tarantula may be even more spectacular than the male.

**Similar Species**

This spider bears a superficial resemblance to the Mexican redrump tarantula. In fact, the species nearly avoided discovery because the only known example was initially dismissed as a small "redrump" and almost not collected. Once the spider was brought back to base camp and studied under good light, the distinctive golden sheen on the legs and carapace was noticed. Although this species and the Mexican redrump have similarly colored abdomens, that of Gutzke's tarantula is more brilliant due to the lack of dark underlying hairs which subdue the intensity of color on the abdomen of *Brachypelma*.

This is the only theraphosid in the Yucatán Peninsula in which mature males lack tibial spurs, making their identification simple and unambiguous.

**Distribution and Habitat**

Known only from the type locality, the west bank of the New River Lagoon on the property of the Lamanai Outpost Lodge, Indian Church Village, Orange Walk District, 17° 45' 08" N, 88° 39' 25" W (Fig. 11.8). The site of collection was a cleared, sloping bank with ornamental plantings and exposed limestone outcrops.

**Abundance**

Gutzke's tarantula must be exceedingly rare. The tarantulas in the immediate area of the type locality have been under intense study since 1994, and well over a thousand individual spiders have been examined, consisting of pygmy, Mexican redrump, and cinnamon tarantulas. To date, only a single specimen of Gutzke's tarantula has been found despite intense searches for more individuals after the presence and appearance of the species became known.

Another explanation, although a remote possibility, can be summoned to explain the rarity of this species. The type locality, as previously mentioned, was a beautifully landscaped garden on the grounds of an eco-lodge. Perhaps the spider arrived here in the soil of some potted ornamental plant that originated in Honduras or Mexico. This would account for the puzzling inability to find additional specimens in such a thoroughly collected area.

**Reproductive Biology**

Nothing is known except that the male was mature when collected on 1 October.

**Natural history**

Due to its rarity, nothing is known of the natural history of this species. The holotype was encountered in the open at 10:30 PM, during a light drizzle.

**Remarks**

*Metriopelma gutzkei* was described as a species of *Crypsidromus* but this genus has since been designated a junior synonym of *Lasiodora* by theraphosid system-

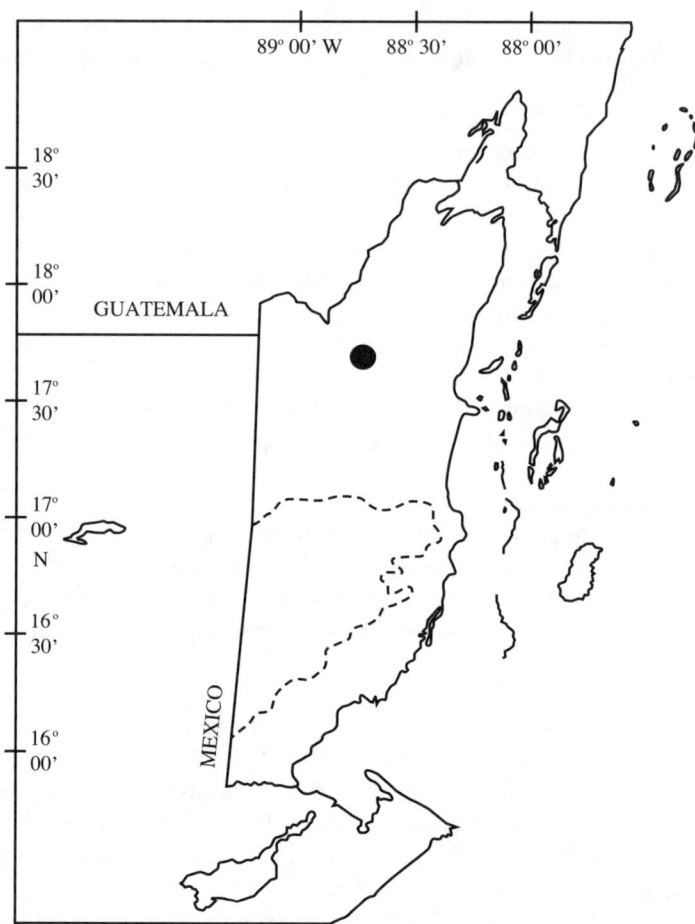

Figure 11.8 Distribution of Gutzke's tarantula.

atist Fernando Pérez-Miles and colleagues. In their comments they indicated that species previously housed in *Crypsidromus* in which the spermatheca was fused should be considered *Metriopelma* species. However, as with *M. gutzkei,* the female of the generic type for *Metriopelma, M. breyeri* Becker, 1878, is unknown. This new assignment cannot be fully resolved for Gutzke's tarantula since the morphology of the spermatheca is unknown. However, Gutzke's tarantula most certainly does not belong in the same genus as the other species that occupy the *Lasiodora* taxon. It is here considered a species of *Metriopelma* but if females are found and the spermatheca should prove to be bilobed the current taxonomic designation would be problematic.

The specific epithet is a patronym in honor of William H.N. Gutzke, a biologist at the University of Memphis and my doctoral advisor.

# ARBOREAL TARANTULAS
## Family Theraphosidae Thorell, 1869
## Subfamily Selenocosmiinae Simon, 1889

**Belizean Species** *Psalmopoeus maya, Psalmopoeus reduncus*

The subfamily Selenocosmiinae is predominantly Old World. Many species are terrestrial, but both Belizean forms are strictly arboreal. A key diagnostic character of both species of selenocosmiine tarantulas in Belize is the presence of a cluster of heavy black bristles, called a lyra, on the maxillae. In common with many other arboreal tarantulas, Belizean selenocosmiines also possess tarsal and metatarsal scopulae which are wider than the tibial segments that precede them (Fig. 11.9), an adaptation to life in the trees.

Most of the world's tarantula species known to be capable of inflicting serious bites on humans are members of this subfamily. Bites may produce worrisome symptoms such as muscle aches and spasms, speech disturbances, and significant local pain. Selenocosmiines are less tolerant of disturbance and should not be handled. There are no documented accounts of bites from the two Belizean selenocosmiine tarantulas, and I have never personally experienced a bite, but caution is advised when dealing with them.

Belizean selenocosmiines are exceptionally agile creatures. When alarmed they may leap into the air and, aided by their feather-like legs, sail softly onto the ground and quickly escape.

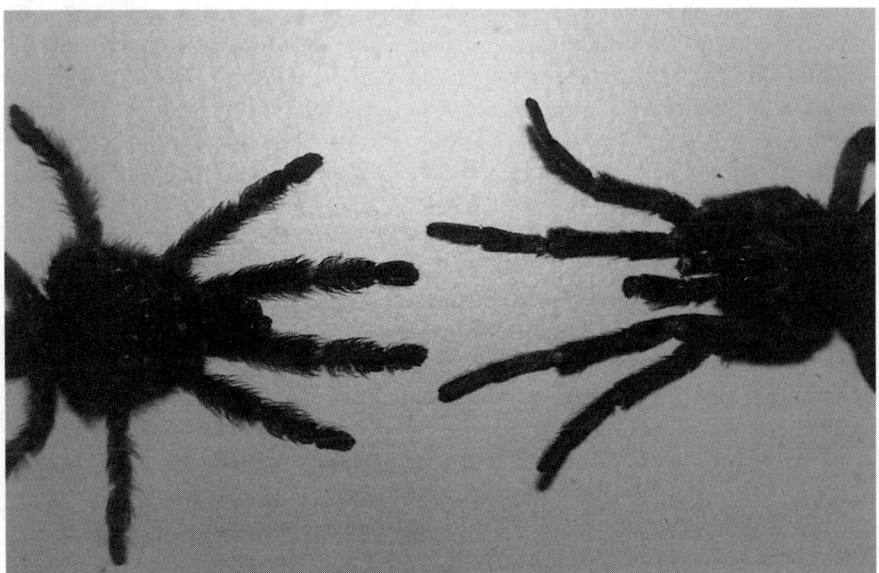

Figure 11.9 Comparison of selenocosmiine (left) and theraphosine (right) tarsi.

## Maya Tarantula *Psalmopoeus maya* Witt, 1996

**Description**

An arboreal tarantula possessing the characteristically broad tarsal and metatarsal scopulae. After a recent molt specimens are overall black (Plate 11.13), but fade to brown (R. West, pers. comm., 2000). A brassy iridescence on the carapace and legs is evident in both sexes. Some reddish brown hairs are visible on legs III and IV. The abdomen has long reddish setae among the underlying black pubescence. Females have a thin white ring around the apex of the four leg joints from femur to tarsus. These rings are particularly evident at the tibia-metatarsus and metatarsus-tarsus joints. I have not seen a male specimen, and no mention of the presence of these rings on males was made in the original species description.

Leg span not recorded for the type specimens, but probably 10–11 cm (3.9–4.3 in), body length 3 cm (1.2 in).

**Similar Species**

The species is defined by its color, which may be darker than other members of the genus. The very similar Costa Rican orangemouth tarantula is grayish brown in overall color as an adult in contrast to this spider which can be nearly black. However, freshly molted subadult Costa Rican orangemouth tarantulas are also nearly black and there is no way to distinguish these from *P. maya*. I have been unable to examine the types as they are not deposited in the University of Hamburg Museum of Zoology (M. Grasshoff, pers. comm., 2000), the original description stating only that the material will be put at the disposal of the museum. Therefore, it is not possible to make informative comparisons between Belizean *Psalmopoeus,* and the only way to distinguish the two species at present is on the very tenuous difference in color. Due to the tendency of color to vary depending on how recently a tarantula has molted the reliability of this distinction as a diagnostic character for *P. maya* is doubtful. The validity of the name *Psalmopoeus maya* remains uncertain and it is possible that the Maya tarantula is simply a dark color variant of the more widely distributed Costa Rican orangemouth tarantula.

**Distribution and Habitat**

Restricted to the type locality, a nameless shallow cave near Sayab Camp in the Slate Creek Preserve, Cayo District, Belize, 300 m elevation, 17° 05' 20" N, 88° 57' 11" W (Fig. 11.10), situated on a hill covered with climax forest. This location is one of the northernmost collection sites for any *Psalmopoeus* species. However, a single specimen of this genus, and possibly also *P. maya,* has recently been found near Uxmal, Yucatán State, Mexico (F. Vol, pers. comm., 2001), so the distribution may be more widespread than presently appreciated. The cave does not give the appearance of being able to support much life. There is no permanent water in the upper levels. Judging from its fully developed eyes, the Maya tarantula is not a troglophilic species. There are unconfirmed reports of spiders which match the description of the Maya tarantula in

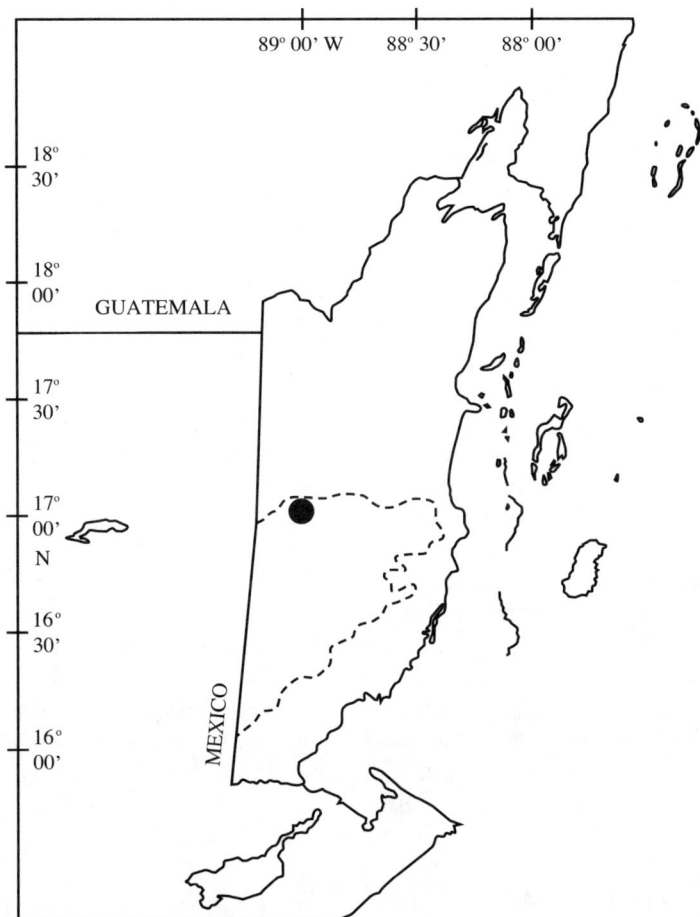

Figure 11.10 Distribution of the Maya tarantula.

other caves along the northern margin of the highlands. At least one of these reports, from arachnologist Stuart Longhorn (pers. comm., 2001), can be considered reliable.

**Abundance**

Unknown. No specimens have been found since the three individuals that were the basis of the original description were collected in 1991.

**Reproductive Biology**

Nothing is known, other than males, which were collected in October, were sexually mature.

## Natural History

Jan Meerman and I explored the collection site in January 1999, but no theraphosids were found, although numerous large, unidentified araneomorph spiders were seen. These spiders could be the source of some of the unconfirmed sightings of Maya tarantulas in caves around the region. The cave is dry but the pocketed limestone walls provide potential refugia for an arboreal tarantula.

The specific epithet refers to the Maya civilization which left its mark everywhere in Belize and permeates life there to this day.

## Costa Rican Orangemouth Tarantula *Psalmopoeus reduncus*
(Karsch, 1880)

### Description

Very similar to the Maya tarantula except lighter in color. Overall grayish brown with long reddish setae on the abdomen and legs (Plate 11.14). Juveniles are almost black. Bright orange hairs are observed around the mouthparts.

Size similar to *P. maya*.

### Similar Species

The wandering spider *Cupiennius salei* is commonly seen on tree trunks and the sides of buildings at night. This spider might be mistaken for an arboreal tarantula due to its large size (leg span up to 14 cm [5.5 in]) and hirsuteness. However, the wandering spider lacks the broad tarsi and metatarsi which are obvious on the tarantula. In addition, the traits that distinguish typical spiders from all mygalomorphs can be used if a close examination is made, but care should be taken because ctenids will bite without much provocation (I can attest to this) and *C. salei* bites are painful. (See also *P. maya*.)

### Distribution and Habitat

The author collected two Costa Rican orangemouth tarantulas in the Toledo District near Blue Creek Village. The species probably occurs throughout southwestern Belize and southeastern Petén to the village of Poptún in the forested karstic limestone hills that border the highlands (Fig. 11.11).

### Abundance

Only two specimens are known from Belize. However, additional specimens were present as indicated by numerous tube webs seen. Most arboreal tarantulas are difficult to find due to the inaccessibility of their retreats. Thus, the Costa Rican orangemouth tarantula may be more abundant in Belize than it appears to be.

### Reproductive Biology

There are no published observations made on Costa Rican orangemouth tarantula reproduction in situ. In captivity, females were observed carrying the cocoon and tending it as terrestrial species do. *Psalmopoeus* characteristically emerge from the cocoon as naked postembryos, which is earlier than in terres-

# Species Accounts

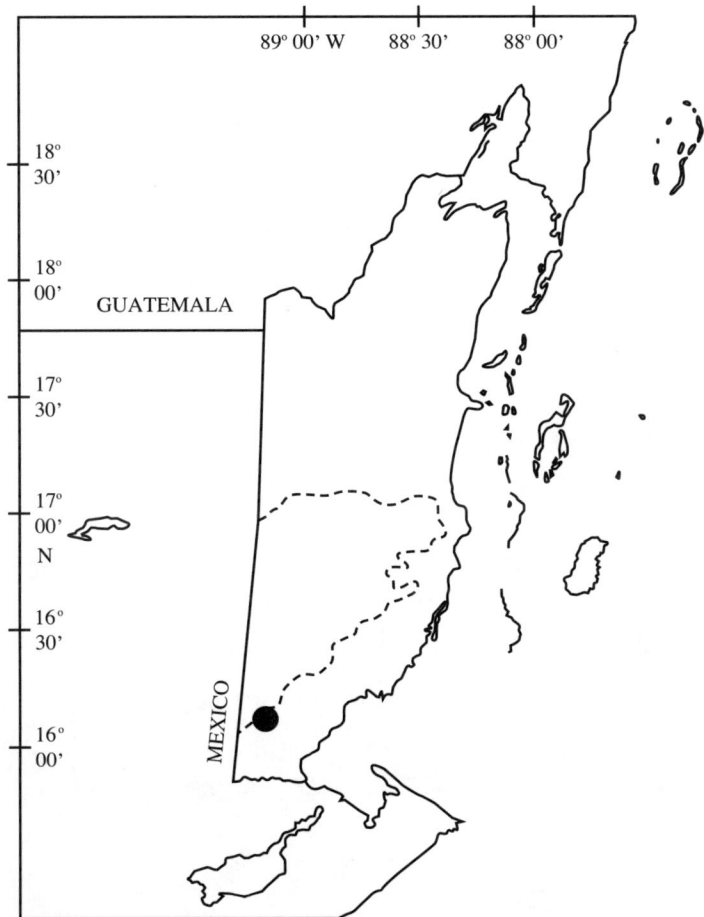

Figure 11.11 Distribution of the Costa Rican orangemouth tarantula in Belize.

trial species. After their first post-emergence ecdysis, the spiderlings exhibit faint chevron markings and a greenish blue sheen on the abdomen (B. Capiz, pers. comm., 2000).

**Natural History**

The microhabitat is typical of other members of the genus. Spiders were found living under corrugated metal roofs, where they had constructed silken tubes. The collection site was in a remnant patch of primary forest which is now an uncommon habitat in Belize.

# Chapter 12

# Extralimital Species

The tarantula fauna of Belize is the most thoroughly studied of any Central American country. In spite of this, there are areas which, due to remote location, might still harbor a few additional forms. In fact, two of Belize's nine known species were discovered in other Central American countries and have only recently been added to the list after being found in difficult-to-access localities. While it is unlikely that species new to science will be discovered, some additional tarantula species already known to occur in Mexico, Honduras, and Guatemala may eventually be found in Belize.

Whenever a species is being considered as an addition to the fauna list of a particular region, it's critically important to base this action on unquestionably sound data. Second-hand information or eyewitness accounts lacking physical proof should serve only as leads for further investigation. For example, a friend in Belize once sent me a series of slides depicting tarantulas from Belize, taken by another individual. Among the portraits of Mexican redrumps and Livingston tarantulas was an image that caused me to stand up and drop the rest of the pile on the floor. In my hands was a slide of a Costa Rican redleg tarantula, *Megaphobema mesomelas,* one of the most beautiful spiders in the world. What shocked me was the fact that this spider was known only from the mountains of Costa Rica, yet here was an image taken in Belize! The note jotted on the slide frame specified that the locale was the Columbia River Forest Reserve in Toledo District, a very remote and mountainous area. I began researching the geological history of Costa Rica and learned that parts of the central mountain range were ancient lands with a genesis similar to the Maya Mountains in Belize. I began to hypothesize that a relict population of *Megaphobema* could exist in Belize, disjunct from the bulk of the range further south. While beginning a rough draft of a formal report of this amazing range extension, which I planned to submit to a scientific journal, I tracked down the photographer, who lived in California, to learn the exact location where the spider had been found. When I spoke to him over the phone, I told him how significant and unexpected this record was. He said he remembered the spider well, but wanted to check his notes to give me the correct information I was seeking. A few days later he called me back with the news that when he found the entry in his field journal, he saw that it was Costa Rica, not Belize, where he had encountered and photographed the tarantula. Apparently he had accidentally included this slide with photos that really were taken in Belize, and my friend had passed these on to me as such without question. Although I was disappointed that such an exciting discovery was bogus, I am forever grateful that I was able to discover the mix-up before I sent a paper off for publication.

The most poorly surveyed location is the extreme southwestern corner, in the hills surrounding the headwaters of the Temash and Pusilhá Rivers. Most of this rugged terrain is roadless. This region is continuous with the highlands that extend westward into Guatemala. It would therefore not be surprising if some species found in eastern Guatemala have a small range into this portion of Belize.

Another area of interest for future species prospecting is the upper elevations of the Maya Mountain massif. Vegetation approaching elfin cloud forest designation is found on some of the peaks along this range, and several new species of frogs have been discovered in these areas (Meyer and Foster, 1996). It is the rule rather than the exception with tarantulas that their geographic distributions tend to be small and constrained by seemingly minor physical or climatic barriers. This may be due in part to the poor dispersal capabilities of these spiders, and because of their intimate relationship with the soil and its specific characteristics. Therefore, endemic species limited to very small areas along the peaks and upper slopes of the Maya Mountains would be in keeping with tarantula biology in general. Some species which have been collected from nearby locations in neighboring countries, and are thus potential additions to the tarantula fauna of Belize, are described below.

## Pygmy chevron tarantula *Acanthopelma rufescens* F.O.P.-Cambridge, 1897

A member of the theraphosid subfamily Ischnocolinae (formerly Acanthopelmiinae), the generic name refers to the spider's "spiny feet," an adornment of stiff spines arranged in a row along the center of the tarsal scopulae. The pygmy chevron tarantula is similar in appearance to *Reichlingia* barychelids and was until recently considered congeneric with them, both genera being tiny spiny-footed mygalomorphs with chevron patterns on their abdomens. The male *A. rufescens* has a simple unmodified embolus and exhibits an unusually blunt tibial spur, although this last character state may simply be a malformation of the single known specimen. The holotype was collected in central Guatemala.

## Rusty brown tarantula *Eurypelma stoica* Chamberlin, 1925

Known only from the holotype, now fragmented and starting to disintegrate (Fig. 12.1), we know little of what this spider looks like in life. However, the palpal emboli remain well preserved and do not resemble anything seen in the genus *Aphonopelma,* to which this specimen has sometimes been assigned. The collection site was the archaeological ruins of Chichén Itza, in Yucatán State, Mexico. This location lies within a drier ecological zone than any in Belize. If the rusty brown tarantula occurs in Belize, it would be in the dry northern scrub near the village of Sarteneja and the Shipstern Nature Reserve.

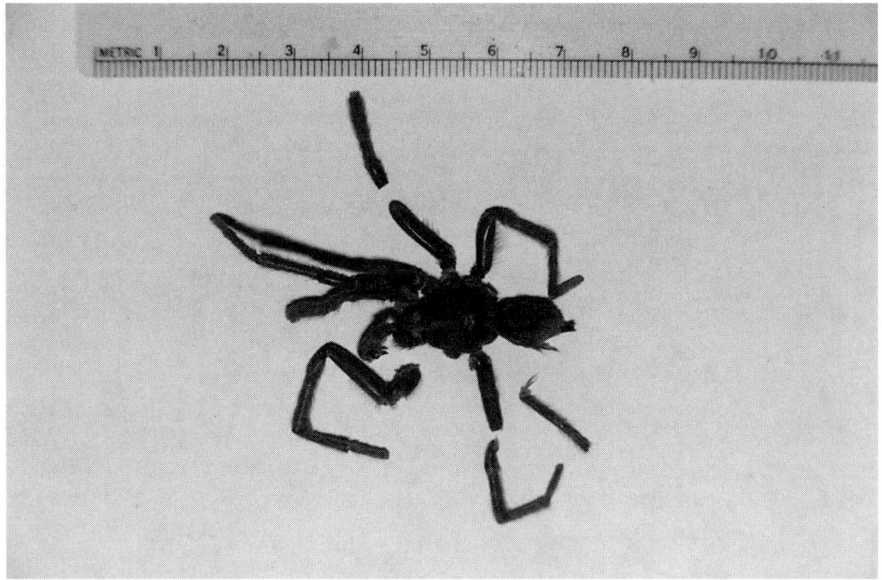

Figure 12.1 Holotype (preserved) of the rusty brown tarantula.

## Guatemalan redrump tarantula *Brachypelma sabulosum* (F.O.P.-Cambridge, 1897)

The Guatemalan redrump tarantula is similar in appearance to the Belizean species, and could be easily overlooked (Fig. 12.2). As discussed in the *B. vagans* species account, this closely related taxon does not penetrate into Belize via the western highlands, as might be expected since this physiographic feature overlaps western Belize and eastern Guatemala. There is a second upland region called the Yalbac Hills, in western Orange Walk District, that is much closer to Guatemalan collection sites, where *B. sabulosum* may occur.

The entire issue regarding the presence of *B. vagans, B. sabulosum,* versus *B. epicureanum* in Belize is one that remains unresolved. This is because all three forms are very similar in superficial appearance, and the extent of intraspecific variation in these spiders is unknown. The point at which a line should be drawn between each species, as opposed to recognizing variation within species, is indefinable at present. A thorough analysis of physical and genetic variation among a large sample of *Brachypelma* from many locations in Belize, Guatemala, and the Mexican Yucatán is required to settle the question. Until this is done there will always be some uncertainty when assigning a species name to any *Brachypelma* found in Belize. In addition to the three described species, there is a possibility that cryptic species have been overlooked by taxonomists due to the inability to discern any distinctions. Comparing DNA sequences among putative populations would reveal such hidden species.

Figure 12.2 Guatemalan redrump tarantula, mature female.

## Mayan orangerump tarantula *Citharacanthus longipes* (F.O.P.-Cambridge, 1897)

This tarantula is recorded as coming from locations in central Guatemala and Tabasco State in Mexico, near the base of the Yucatán Peninsula. Confirmed examples of this species have not been seen in over 100 years. Due to the poor condition of the preserved specimens and the incompleteness of the original description it is difficult to state characteristics that could distinguish this species from the Cayo or Livingston tarantulas under field conditions. The types lack plumose setae on femorae I and II, which is diagnostic for the other Yucatecan *Citharacanthus* tarantulas. On Cayo and Livingston tarantulas, these setae impart a velvety sheen that is visible with the naked eye.

# Chapter 13

# Beyond Belize

Scientific studies which shed light on the broadest issues are the most valuable. This is why the best way to devise a course of biological study is to identify the question of interest first, and then decide what organism, in what part of the world, is best suited to answer that question. By following this procedure, no matter how obscure the organism or how remote the study site, the finished work is bound to offer insights into areas beyond the immediate subject. Some biologists tend to orient toward a particular organism rather than a biological question, and conceive their research projects by asking what new question could be asked about that taxon. Although species-specific research helps fill in missing natural history details, the findings are only applicable to the animal that was studied, and are more broadly relevant only in rare instances or under very narrow circumstances. As a result of the difference between these two approaches, information which at first glance appears arcane may indeed be so, or may actually be useful for understanding much broader topics.

Why study tarantulas, and why focus on Belize? I began working in Belize because it is a natural laboratory for learning about tarantula biology. A considerable ecological diversity is packed into a tiny geographic area and the small scale of in-country travel coupled with good roads make the spiders easily accessible. Belize is full of tarantulas, often living crowded closely together, making large study samples possible. Scientific conclusions are based on statistical analysis of large numbers of observations. Robust results are easier to achieve in Belize than in many other regions where tarantulas occur. With its variety of habitats and an abundant supply of spiders representing an interesting array of species, Belize presents a perfect opportunity to explore the mysteries of tarantula biology.

But what can the tarantulas of Belize tell us about other tarantulas, or about spiders in general? Can the study of Belizean tarantulas contribute to our understanding of neotropical ecology and the conservation issues that impact it?

## Historical Biogeography

One of the most far reaching aspects of Belizean tarantula natural history is the connection between their zoogeography and the biogeographical history of Central America. The isthmus is a young land, long predated by North and South America. These two continents were isolated land masses for most of their existence and their faunas evolved separately. Over the past 50 million years, it is thought that a chain of islands was intermittently present between North and South America and that several times these islands congealed into one continuous land bridge—the Central American isthmus—and connected the two continents. Shifting crustal plates and the resulting geological upheavals along with rising and falling sea levels created an everchanging Central America—unbroken at times but usually with many disconnections that kept the two faunas evolving independently from each other. Whenever the two continents were connected, some elements of the disparate faunas were able to disperse north or south into new territory, only to be isolated again as the isthmus was broken. Propelled by superior dispersal capability and the ability to adapt to new environments, certain organisms were able to establish themselves in the new territory. Other species, either because they were unable to traverse the entire length of Central America or were halted by unfavorable environmental conditions along the way, never spread far beyond their native continent and remain restricted to it today. Thus South America has the arboreal *Avicularia* tarantulas, with only one species straying northward to eastern Costa Rica (*A. avicularia*), and the genus *Brachypelma* has evolved into a great variety of forms in Mexico and Central America yet stops abruptly near the South American border. The present distribution of North, Central, and South American life forms is partially the consequence of a prehistoric exchange of species and their subsequent fates.

Just as regional compositions of human ethnic ancestry in the United States are a reflection of historical social trends and the waves of immigration they elicited from particular regions, the present day distribution of animal species in Central America offers clues to their histories, and thus, to the origin of the land itself. By focusing on a specific faunal element, a less complicated story is delineated and patterns become more discernible. Looking at tarantulas in this way, one of the most productive aspects to begin pondering is a comparison between taxa which are present versus those that are expected but absent. In this way, hypotheses regarding the prehistoric physiography of the region can be developed and tested.

The genus *Brachypelma* is abundant in Belize, as would be expected due to the numerous species occurring both north and south. Yet the apparent absence of another widespread New World tarantula genus, *Aphonopelma,* is unexpected and puzzling. *Aphonopelma* species are found from the United States and throughout Mexico, and its distribution approaches Belize from the north at least as close as Yucatán State. Two species of the genus are known from localities south of Belize, *A. seemanni* (Nicaragua and Costa Rica) and *A. burica* (Costa Rica). Thus Belize is surrounded by *Aphonopelma* tarantulas but none are known from within its border. Why would these two genera, *Brachypelma* and *Aphonopelma,* both having traversed the Central American land bridge, exhibit such different distributions with regard to Belize? The likely evolutionary origin of both genera is Mexico, with subsequent radiation south, but perhaps during widely separated periods of history.

Three hypotheses can be proposed. The perception that *Aphonopelma* does not occur in Belize may be incorrect, and there may well be a species that remains undiscovered. This seems unlikely given how well-collected the country is, yet there are still a few ecologically distinct regions that require more exploration. A second possibility is that *Aphonopelma* may have established its presence in Belize but is now represented by an evolutionary descendent. A suspect taxon is the cinnamon tarantula, which is restricted to Belize and adjacent areas and has no known related species outside this region. The cinnamon tarantula may be the echo of an earlier presence of primitive *Aphonopelma* stock, now evolved into a distinct form. Merely speculative at the moment, this conjecture could be tested with DNA analysis. If the cinnamon tarantula or another Belizean species does not turn out to be a sister taxon of *Aphonopelma* and the absence in Belize is real, it would suggest that something unique occurred during the formation of the land that is Belize. Why then, did the ancestral *Aphonopelma* disperse southward from Mexico, reach points farther south down the isthmus, but fail to establish in Belize? Was part of Belize isolated by ocean during a period when the land bridge was intact as far south as Costa Rica, and this was the time that *Aphonopelma* dispersed southward, thus bypassing Belize? Answers to such questions await future research and in any case are beyond the scope of this book, but they serve to illustrate how the study of a seemingly narrow subject, such as the tarantulas living in a small Central American country, can generate bigger questions and help provide answers to a range of broader topics.

## Niche Partitioning

Another topic, this one ecological, that the study of Belize's tarantulas can contribute to is niche partitioning and the competitive struggle between sympatric species. Fundamental to an understanding of how species assemblages interact is knowing how they divide the dimensions of their shared environment—spatially, temporally, and behaviorally—so that conflicts are reduced or avoided. For instance, some animals are active at night while their potential competitors are more likely to restrict their movements to the daylight hours. When potential competitors are active at the same time, competition for the same resource may be reduced by spatial separation, as in the case of the Livingston and Costa Rican orangemouth tarantulas. Both are active at night and prey on insects, but the Livingston tarantula does so at the entrance of its burrow on the forest floor, and feeds on terrestrial insects, while the foraging Costa Rican orangemouth tarantula may be only steps away, but resting on the side of a tree trunk and catching arboreal katydids or crickets.

In Belize, there are several multi-species assemblages of tarantulas that are typical: Mexican redrump, cinnamon, and pygmy tarantulas in the northern lowlands, Mexican redrump and Cayo tarantulas in the highlands, and Mexican redrump, Livingston, and Costa Rican orangemouth tarantulas in the southern Maya Mountain foothills. Each of these sympatric groupings offers ways to test resource partitioning theories and by doing so, scientists will shed light on the mechanisms of tropical ecology in general.

## Invasive Species

The impact of invasive species is of great concern to conservation biologists. Many ecological disasters have been caused by the accidental or all too often, the intentional introduction of an exotic species into a fragile ecosystem. Tarantulas have not been spared this fate, the extinction of the Jamaican tarantula due to the introduction of the mongoose being one sad example. Whether it's brown tree snakes systematically decimating the birds of Guam, rabbits consuming native plants on Round Island and causing the collapse of the entire ecosystem, mongooses exterminating unique and beautiful species of snakes and tarantulas in the West Indies, or the pestilence of rats virtually anywhere they've been introduced, an invader that gains a foothold on foreign soil does so at the expense of some competitively weaker native species.

In 1996, Mexican redrump tarantulas were discovered in Florida. About 30 spiders were living in a citrus grove near Fort Pierce in St. Lucie County. The spiders had apparently been present for a considerable time because they had constructed burrows. The source of the introduction, the ability of the spiders to survive over the long term, and whether the tarantulas were part of a larger established population living in the area, is not known. No other aggregations were found, and state agricultural officials destroyed the ones they discovered. One of the concerns about this introduction was the possible impact on native wolf spiders, which have similar habitat requirements and would probably find young tarantulas to be competitors for the same limited component of insect prey. Although wolf spiders are sympatric with Mexican redrump tarantulas in Belize, this association has evolved under a different set of conditions. The wolf spiders of Florida, some of them endemics with very narrow habitat requirements, have never faced a tarantula competitor. Mexican redrump tarantulas are a beautiful member of the Belizean wildlife community and should be cherished as such. They have no place in Florida. The action by officials to eradicate the tarantulas, considering the impact they may have had on the native ecosystem, was completely appropriate.

## Human Impact on Wildlife

The preceding example of a potentially detrimental effect of tarantulas on the environment is exceptional. In practically every instance, it is the activities of man that destroys ecosystems, and in the tropics, tarantulas are one of the many casualties. Belize's tarantulas hold answers to the urgent questions that conservation biologists are asking regarding the rapidly vanishing forests of Central America.

The tropical forests of Belize face ever increasing pressure from people seeking land for homes and crops. Corporations prospecting for wealth put wild areas at further risk. Meanwhile, conservationists search for ways to distinguish landscapes that are in critical distress versus those where human activity has not drastically upset the natural balance. Areas where man and nature coexist are particularly interesting because the pragmatic truth about conservation is that it must include some compromise, and strike a balance between the interests of modern society and nature. Most wild areas will receive some degree of protection only if such compromises are reached.

Preliminary research suggests that tarantulas are excellent bio-indicators of ecosystem health. Tarantulas, like most arthropods, are highly sensitive to pesticides that build up in the soil or insinuate into the trophic web. Unlike many arthropods, tarantulas are sedentary and therefore easy to locate and monitor as individuals. Tarantulas also hold the distinction of being exceptionally long-lived for an arthropod, so the effects of long-term exposure to environmental toxins can be studied in individual animals, making research modeled on tarantulas much more relevant to vertebrates—including humans—than information gathered from other types of spiders or insects.

Studies comparing tarantula population densities in Belizean milpas of varying size and age are underway, and will help to define the parameters within which slash and burn agriculture might coexist with natural ecosystems. Tarantulas are noticeably abundant in small clearings when these are imbedded in an overall matrix of forest or scrub. Species adapted to open habitat such as Mexican redrump, cinnamon, and pygmy tarantulas are more common in clearings than in the surrounding forest. However, the presence of nearby forest is still critical to the health of the population, as it probably serves as the primary source of insect prey. In regions where the land has been extensively cleared and no forest buffers remain, tarantula densities decline over time. For several years in one particular milpa in Orange Walk District, it was always possible to find 100 or more burrows of Mexican redrump and cinnamon tarantulas. At the time, the milpa was a small plot planted with banana and plantain, and was situated within a large expanse of forest and scrub. A sudden increase in the nearby human population (due to the influx of refugees fleeing the violence of civil unrest elsewhere in Central America) led to a dramatic change in the character of the landscape. Within only a few years the entire countryside was denuded of forest as a continuous patchwork of active agricultural fields was established (Fig. 13.1). Although both Mexican redrump and cinnamon tarantulas can still be found, they are far less abundant and are distributed sparsely in widely scattered burrows. A recent search at the same site where more than 100 tarantulas could be found only 5 years earlier turned up only a dozen large adult "redrumps." No juveniles were present, suggesting the dwindling Mexican redrump population was reproductively senescent, and the cinnamon tarantulas had vanished altogether. As tropical deforestation proceeds, research on Belize's tarantulas may help conservationists throughout the neotropics devise ways to manage the land. Hopefully, we can strike a balance between the welfare of humans and the preservation of wildlife.

## The Future

Long heralded as the most conservation conscious and unspoiled Central American country, Belize is starting to face the same pressures that have largely destroyed the natural areas throughout the rest of the region. It is still true that the people of Belize, and their government, maintain an environmental ethic that surpasses most countries, including nations which have far more resources to devote to domestic conservation issues. However, a contributing factor that has spared the Belizean forests and wetlands from development is the relatively small human pop-

Figure 13.1 Tropical forest cleared by slash and burn.

ulation—the smallest in Central America. Most Belizeans live and work in the vicinity of a handful of towns and villages, further reducing pressure on the natural landscape. Inevitably, Belize's population is beginning to grow at an increasing rate, and population centers are expanding into once empty land.

The first impact of these changes on tarantulas has been on forest-adapted species. As the forests become fragmented and shrink, so do the distributions of Livingston and Gutzke's tarantulas. Once whole districts are deforested by migratory-destructive slash and burn, even the most resilient species will disappear. Some people may not find this troublesome, but all life on earth is interconnected and eventually, as the tarantulas go so do we. I urge everyone to develop a personal conservation ethic that includes, if not affection, then at least a tolerance of every living creature, including the misunderstood and under appreciated tarantulas of Belize.

# Gazetteer

Locality records are presented for all specimens examined during research for this book. Collection abbreviations: AMNH–American Museum of Natural History, New York. BMNH–British Museum of Natural History, London. CFF–Carol Farneti Foster, Belmopan, Belize. FMNH–Field Museum of Natural History, Chicago. JCM–Jan C. Meerman, San Antonio, Belize. MCZ–Museum of Comparative Zoology, Harvard University, Cambridge, Massachusetts. MLH–Mark L. Howells, Lamanai Field Research Center, Indian Church Village, Belize. RCW–Rick C. West, Victoria. SBR–Steven B. Reichling. SMF–Natur-Museum und Forschungsinstitut Senckenberg, Frankfurt.

*Acanthopelma rufescens:* GUATEMALA: specific location unrecorded, holotype (BMNH).

*Brachypelma epicureanum:* MEXICO: Yucatán: Chichén Itza, paratype (MCZ); specific location unrecorded (SBR).

*Brachypelma sabulosum:* GUATEMALA: specific locality unrecorded (SBR).

*Brachypelma vagans:* BELIZE: Ambergris Cay: 11.3 km N San Pedro Town (examined and released, SBR); 8.0 km S San Pedro Town (examined and released, SBR). Belize District: Belize Zoo and Tropical Education Center (CFF); Maskall Village (AMNH). Cayo District: Forest Drive, Belmopan (CFF); Guanacaste National Park (CFF); Junction Cristo Rey Road and Chiquibul Road (AMNH); Mile 34, Western Highway (CFF); 2.0 km SE San Ignacio Town (AMNH); Las Cuevas Research Station, Chiquibul Forest Reserve (examined and released, SBR). Corozal District: Santa Rita Archeological site (AMNH); 0.8 km W Sarteneja Village (AMNH); Shipstern Nature Preserve (AMNH). Orange Walk District: Dawson's Creek (AMNH); Gallon Jug (CFF); Lamanai Outpost Lodge, Indian Church Village (AMNH); Las Milpas, Rio Bravo Conservation Area (AMNH); Shipyard (AMNH). Stann Creek District: Bocawina Hill (FMNH); Cockscomb Basin Wildlife Sanctuary, picnic area (examined and released, SBR); 4.8 km W Hopkins Village (examined and released, SBR); Middlesex Village (FMNH). Toledo District: Belize Foundation for Research and Environmental Education, Bladen Nature Preserve (AMNH); Blue Creek Village (examined and released, SBR); Joe Taylor Creek (AMNH). Turneffe Islands: Blackbird Cay (MLH); Calabash Cay (MLH). MEXICO: Yucatán Peninsula, holotype and paratype (BMNH).

*Citharacanthus livingstoni:* BELIZE: Cayo District: Cockroach Camp (AMNH). Toledo District: Belize Foundation for Research and Environmental Education, Bladen Nature Preserve (AMNH); Blue Creek Rain Forest Preserve (AMNH); Doyle's Delight (leg. S. Matola, slide examined, SBR); Union Camp (AMNH). GUATEMALA: Livingston, holotype (SMF).

*Citharacanthus meermani:* BELIZE: Cayo District: Caracol (AMNH); Hidden Valley Inn, Mountain Pine Ridge (AMNH); 2.0 km SE San Ignacio Town, holotype (AMNH); Slate Creek Preserve (AMNH). Stann Creek District: Middlesex Village (FMNH); Las Cuevas Research Station, Chiquibul Forest Reserve (examined and released, SBR).

*Crassicrus lamanai:* BELIZE: Belize District: Altun Ha (examined and released, SBR); Belize Zoo and Tropical Education Center vicinity (Foster residence) (CFF); Maskall Village (examined and released, SBR). Cayo District: Belmopan (JCM); 5.0 km S Belmopan, Hummingbird Highway (RCW); 2.0 km SE San Ignacio Town (AMNH). Orange Walk District: Lamanai Outpost Lodge, Indian Church Village (AMNH); 0.5 km W New River Lagoon, holotype and paratypes (AMNH); Orange Walk Town (examined and released, SBR). Stann Creek District: 4.8 km W Hopkins Village (examined and released, SBR). Corozal District: Corozal Town (MLH). Toledo District: Belize Foundation for Research and Environmental Education, Bladen Nature Preserve (AMNH); Blue Creek Village (examined and released, SBR).

*Eurypelma stoica:* MEXICO: Yucatán: Chichén Itza, holotype (MCZ).

*Metriopelma breyeri:* MEXICO: Guanajuato (BMNH).

*Metriopelma gutzkei:* BELIZE: Orange Walk District: 0.1 km W New River Lagoon, holotype (AMNH).

*Psalmopoeus maya:* GUATEMALA: Izabal: Sierra de Santa Cruz (slide examined, courtesy Rick C. West).

*Psalmopoeus reduncus:* BELIZE: Toledo District: Blue Creek Rain Forest Preserve (AMNH).

*Reichlingia annae:* BELIZE: Belize District: Belize Zoo and Tropical Education Center, mile 29 Western Highway, (AMNH); 12.9 km S jct. Northern Highway and Old Northern Highway (AMNH). Orange Walk District: August Pine Ridge (AMNH); Dawson's Creek savanna, 2 km E New River Lagoon (AMNH); 0.5 km W New River Lagoon, Indian Church Village, holotype and paratypes (BMNH); Lamanai Outpost Lodge (AMNH).

*Reichlingia* sp: BELIZE: Cayo District: 2.0 km SSE San Ignacio Town (AMNH); 2.4 km SE San Ignacio town (AMNH).

# Bibliography

Ausserer, A. 1875. Zweiter Beiträg zur Kenntnis der Arachniden-Familie der Territelariae Thorell (Mygalidae Autor). *Verhandl. K. K. Zool.-Bot. Gesell. Wien* 25:125–206.
Bertani, R. 2000. Male palpal bulbs and homologous features in Theraphosinae (Araneae, Theraphosidae). *J. Arachnol.* 28:29–42.
Cambridge, F.O.P.- 1897. Arachnida-Araneida. In *Biologia Centrali-Americana,* Godman, F.D. and O. Salvin, 2:1–40.
Chamberlin, R.V. 1925. New North American spiders. *Proc. California Acad. of Sci.* 14:105–142.
Cooke, J.A.L., V.D. Roth, and F.H. Miller. 1972. The urticating hairs of theraphosid spiders. *Am. Mus. Novit.* 2498:1–43.
Hartshorn, G., L. Nicolait, L. Hartshorn, G. Bevier, R. Brightman, J. Cal, A. Cawich, W. Davidson, R. Dubois, C. Dyer, J. Gibson, W. Hawley, J. Leonard, R. Nicolait, D. Weyer, H. White, and C. Wright. 1984. *Belize: country environmental profile: a field study.* Robert Nicolait and Assoc., Ltd., Belize City.
Karsch, F. 1880. Zur arachnidengatfung *Theraphosa. Zeitschr. Gesell. Naturwiss.* 53:843–846.
Marshall, S.D. 1996. *Tarantulas and Other Arachnids.* Barron's Educ. Ser., Hauppauge, NY.
Marshall, S.D. and G.W. Uetz. 1990. Incorporation of urticating hairs into silk: a novel defense mechanism in two neotropical tarantulas (Araneae, Theraphosidae). *J. Arachnol.* 18:143–149.
Marshall, S.D. and G.W. Uetz. 1993. The growth and maturation of a giant spider: *Theraphosa blondi* (Latreille, 1804) (Araneae, Theraphosidae). *Revue Arachnologique* 10:93–103.
Marshall, S.D., and G.W. Uetz. 1990. The pedipalpal brush of *Ephebopus* sp. (Araneae, Theraphosidae): evidence of a new site for urticating hairs. *Bull. Br. Arachnol. Soc.* 8:122–124.
Meyer, J.R. and C.F. Foster. 1996. *A Guide to the Frogs and Toads of Belize.* Krieger Publishing Co, Malabar, FL.
Pérez-Miles, F. 1998. Notes on the systematics of the little known theraphosid spider *Hemirrhagus cervinus,* with a description of a new type of urticating hair. *J. Arachnol.* 26:120–123.
Pérez-Miles, F., S.M. Lucas, P.I. da Silva Jr., and R. Bertani. 1996. Systematic revision and cladistic analysis of Theraphosinae. *Mygalomorph* 1:33–68.
Platnick, N.I. 2000. *The World Spider Catalog.* Am. Mus. Nat. Hist., New York.

Raven, R.J. 1985. The spider infraorder Mygalomorphae (Araneae): cladistics and systematics. *Bull. Am. Mus. Nat. Hist.* 182:1–175.
Reichling, S.B. 2001. Theraphosid spiders surveyed at Las Cuevas. *Newsl. Las Cuevas Res. Sta., Belize* 8:6.
Reichling, S.B. 2000. Group dispersal in juvenile *Brachypelma vagans* (Araneae, Theraphosidae). *J. Arachnol.* 28:248–250.
Reichling, S.B. 1999. Nearest neighbor relationships among tarantulas in Belize. *Southwestern Nat.* 44:518–521.
Reichling, S.B. 1997. A diminutive new species of *Acanthopelma* from Belize (Araneae: Theraphosidae). *Bull. Br. Arachnol. Soc.* 10:337–340.
Reichling, S.B. 1997. A new species of *Crypsidromus* from Belize (Araneae, Mygalomorphae, Theraphosidae). *J. Arachnol.* 25:49–52.
Reichling, S.B. 1996. The role of incubation temperature and food intake on phenotype of terrestrial theraphosid spiders (Araneae, Mygalomorphae) [PhD dissertation]. Memphis (TN):University of Memphis. 174 pp. Available from University Microfilms, Ann Arbor, MI: 9730646, 1194 MEMP-D.
Reichling, S.B., and C. Tabaka. 2001. A technique for individually identifying tarantulas using passive integrated transponders. *J. Arachnol.* 29:117–118.
Reichling, S.B., and R.C. West. 2000. A new species of tarantula spider (Araneae, Mygalomorphae) from the Cayo District of Belize. *Southwestern Nat.* 45:128–132.
Reichling, S.B., and R.C. West. 1996. A new genus and species of theraphosid spider from Belize (Araneae, Theraphosidae). *J. Arachnol.* 24:254–261.
Rudloff, J.-P. 2001. Anmerkungen zur systematischen Stellung von *Acanthopelma rufescens* F.O.P.-Cambridge, 1897 und *Acanthopelma annae* Reichling, 1997 (Ischnocolinae: Theraphosidae: Mygalomorphae), sowie die Einrichtung einer neuen Gattung *Reichlingia* gen. nov. (Mygalomorphae: Barychelidae: Trichopelmatinae). *Arthropoda* 9:14–20.
Schmidt, G., and D. Weinmann. 1996. Eine weitere *Citharacanthus*-species aus Guatemala *Citharacanthus livingstoni* sp. n. (Araneida: Theraphosidae: Theraphosinae). *Arachnologisch Mag.* 4(3):1–11.
Valerio, C.E. 1979. Arañas terafósidas de Costa Rica (Araneae: Theraphosidae), 2. *Psalmopoeus reduncus,* redescription, distribución y el problema de dispersión en terafósidas. *Revista Biol. Trop.* 27:301–308.
Smith, A.M. 1994. *Theraphosid Spiders of the New World, Vol. 2, Tarantulas of the USA and Mexico.* Fitzgerald Publishing, London.
Straughan, R.P.L. 1975. *Adventure in Belize.* A.S. Barnes and Co., New York.
Witt, G. 1996. Eine neue *Psalmopoeus*-Species aus Belize *Psalmopoeus maya* sp. n. (Araneida: Theraphosidae: Aviculariinae). *Arachnologisch Mag.* 4:1–14.

# Glossary

**apical**   Located at the outermost tip.
**apophysis** (*pl.* apophyses)   A projection of the exoskeleton.
**arboreal**   Living in trees.
**basoprolateral**   The inner surface, as on the leg, near the junction with the body (baso-) and facing the body when the leg is folded naturally (-prolateral).
**bauplan**   A basic set of physiological and morphological parameters that are necessary and within which an organism can be formed.
**bifurcate**   Forked with two branches.
**capitate**   Distinctly clubbed.
**chelicera** (*pl.* chelicerae)   The muscular segment to which the fang attaches and used to manipulate the fang and serve as an opposing surface against which prey is masticated.
**clavate**   Distinctly broad and flattened.
**cymbium**   The highly modified terminal segment of the pedipalp which appears after the ultimate ecdysis in male tarantulas, on which the palpal bulb is attached.
**ecdysis**   The shedding of the exoskeleton.
**elytron** (*pl.* elytra)   The modified forewing of beetles which serves as a protective covering over the hindwing.
**epigastric furrow**   A fold of the exoskeleton on the anterior underside of the abdomen, visible as a slit, beyond which lies the opening to the reproductive organs.
**exuvium** (*pl.* exuvia)   A cast-off exoskeleton.
**fossorial**   Living underground, a burrowing animal.
**holotype**   The specimen that was used as the standard for the species in the original scientific description.
**incrassate**   Swollen.
**lyra**   A patch of stiff, bristle-like setae located on the prolateral surface of the maxillae in the two arboreal tarantulas of Belize.
**maxilla** (*pl.* maxillae)   The basal segment of the first leg-like appendage (pedipalp), lying on either side of the tarantula's oral opening.
**monophyletic**   Pertaining to a group of organisms believed to share a recent common ancestor and representing a single evolutionary lineage (see polyphyletic).
**obligative cavernicolic**   Compelled to living inside caves.
**oophagy**   The practice of consuming eggs. For spiders this normally refers to eating eggs within one's own clutch.

**paratype**  An additional specimen referred to in the paper describing the species and used in its description.

**plesiomorphic**  Of primitive form; used when describing physical characters which are believed to be ancestral.

**polyphyletic**  Pertaining to a group of organisms believed to have derived from distantly related ancestors and thus representing two or more evolutionary lineages (see monophyletic).

**postembryo**  Naked, unpigmented juvenile tarantulas between the first ecdysis, performed inside the cocoon, and the second molt when the setae and coloration appear.

**prolateral**  The inner surface, as on the surface of a spider's leg which faces the body when the leg is naturally folded against it.

**pubescence**  The short, dense undercoat of setae clothing the legs and body of most tarantulas.

**seta** (*pl.* setae)  A hairlike extension of the exoskeleton comprising the furry covering on the body of many arthropods.

**scopula** (*pl.* scopulae)  Specialized setae on the ventral surface of the tarsi, used for gripping smooth surfaces when the tarantula climbs.

**scopular pad**  A dense concentration of many scopulae, usually located on the ventral surface of the tarsi.

**spermatheca** (*pl.* spermathecae)  Pocketlike receptacles composed of chitin which reside internally in the tarantula's abdomen and into which seminal fluid is deposited and retained.

**spinose**  Resembling a spine.

**spiracle**  The external opening to the tracheal system in arthropods.

**stridulation**  The production of sound by rubbing two surfaces together.

**sympatric**  When the distribution of two or more different kinds of organisms overlap in the same geographical area.

**trichobothria**  Specialized and highly modified setae that detect minute air currents and serve as sensitive motion detectors.

**troglophilic**  Fond of living in caves, but not limited to doing so.

**urticaceous seta**  A hairlike structure with barbs that causes an irritation when contacting skin, and used by many tarantulas as their primary means of defense against some predators and parasites.

# Index

Numbers in **boldface** indicate photographs.

**A**
*Acanthopelma,* 18, 85, 88, 110
*Acanthopelma rufescens,* 86, 110
Age of Mygalomorphs, 17
Aggregative dispersal:
   discovery of, 9–10
Antelope spider, 100
*Aphonopelma:*
   absence in Belize, 114–115
   Arkansas, 3
   lifespan, 57
*Aphonopelma crinitum,* 24–25
*Atrax robustus. See* Sydney
   funnelweb spider.
*Atterocopus fimbriunguis,* 16
Ausserer, Anton, 12

**B**
Baerg, William J.:
   in Jamaica, 1
   life history studies by, 57
   observations on *Aphonopelma*
   *crinitum,* 24–25
Barychelidae, 18, 85, 110
Bites, 56–57, 67, 90, 103
   allergic reaction to, 57
Book lungs, 16
*Brachypelma:*
   all-black species, 4, 7, 92
   Florida introduction of, 116
   systematics, 111
*Brachypelma auratum. See*
   Mexican flameknee tarantula.
*Brachypelma smithi. See*
   Mexican redknee tarantula.
Burrows:
   abandoned, 64–65
   aggregations, 40, **41,** 92, 97, 99,
   depth, 39
   entrance veil, 39–40, **40,** 62
   marking of, 65

   plugging, 48–50, **49,** 88
   wolf spider, 40, 63–64, **64**

**C**
Caretaker of the Forest, 7
Chamberlin, Ralph Vary, 12
CITES, 23–24, 70
Collecting:
   justification for, 67, 69
   permits, 69–70
Courtship, 44–47
Ctenidae, 15
Ctenizidae, 17
*Cupiennius salei,* 106
*Cupiennius* sp., **16**

**D**
Dipluridae, 17
Dispersal, 50
   aggregative, 9–10, 51–52, **52,** 93

**E**
Eggs, 48
Embolus:
   function of, 44–45
   morphology, 77, **78**
Exportation:
   permit, 70
   procedure, 70–72
Eye shine, 64

**F**
Facultative burrowers, 37
Fangs, 16–17, **17**
"Fishing" technique, **66,** 66–67

**G**
Geology:
   formations, 11
   history, 113–114
Gutzke, William H.N., 102

## H
*Hogna carolinensis,* **15,** 29
Howells, Colin, 4–6
Howells, Mark, 4–7, 10
Howells, Monique, 5–6, 10

## I
Ischnocolinae, 18
*Ischnocolinopsis acutus,* 17

## J
Janowski-Bell, Margaret E., 43
Juveniles:
    Costa Rican orangemouth tarantula, 107
    emergence from cocoon, 51
    feeding study, 2
    kin recognition, 51
    Mexican redrump tarantula, 91
    passive maternal feeding of, 50
    pygmy tarantula, **49**
    retreats, 37, **39,** 69, 94
    urticating hair patch, 55

## K
Kotzman, Mandy, 50

## L
Labyrinth of the Tarantulas, 62
Lacey Act, 23–24
Lamanai Archaeological Reserve, 9, 13, 26–27, **27**
Leaf-cutter ants, 98
Legs:
    *Crassicrus,* 100
    segments, 73, **75**
Leopold, Aldo, 20–21
Lifespan, 21
    comparative, 57
    male, 47
Longhorn, Stuart, 105
Lycosidae, 15

## M
Maternal feeding, 50
Marshall, Sam, 37, 57
Mating season, 41–43, 92–93
Matola, Sharon, 94
Meerman, Jan, 98, 106

*Megaphobema mesomelas,* 109
Mesothelae, 15
Mexican flameknee tarantula, 24
Mexican redknee tarantula:
    CITES protection of, 23
    importation, 24

## O
Obligate burrowers, 37
Oophagy, 51
Opisthothelae, 15

## P
Pérez-Miles, Fernando, 102
Pickard-Cambridge, Frederick, 12
Prentice, Tom, 77

## R
Raven, Robert J., 18
Reichling, Ann, 2, 3, 88
Rio Bravo Escarpment, 35, 99
*Rosamygale,* 17

## S
*Sand County Almanac,* 20
Slash and burn, **118**
    impact of, 25–28, 34, 117–118
    traditional, 25–26, **26**
    migratory destructive, 26–27, **27**
Sperm storage, 47
Sperm web, 43–44, **44,** 46
Spermathecae, **47**
    comparative morphology, **78,** 78–79
    function of, 47
Spinnerets, 16, 76
Sydney funnelweb spider, 17

## T
Tarantula hawk, 54–55
Tarsal claws, 74
Theridiidae, 15
Tibial spurs:
    comparative morphology, 77–78, **78**
    use of, 45
Tikal, 13, 92
Tracheal system, 16
Thoreau, Henry David, 20
Trichobothria, 85
Trichopelmatinae, 18

Trip lines, 53
Trophic eggs, 51
Tube web, 39, **40,** 68, 106, 107

**U**
Urticating hairs:
  irritating capabilities, 56
  types, 55
  use, 44, **56** 91, 100

**W**
*Walden,* 20
Wetlands, 29, 95
Wildlife Protection Act, 69
Witt, Gunther, 13

**Y**
Yalbac Hills, 111